Reasoning with Concepts

Reasoning with Concepts

Conceptual Spaces as a Framework

Peter Gärdenfors and Matías Osta-Vélez

The MIT Press
Cambridge, Massachusetts
London, England

The MIT Press
Massachusetts Institute of Technology
77 Massachusetts Avenue
Cambridge, MA 02139
mitpress.mit.edu

The MIT Press would like to thank the anonymous peer reviewers who provided comments on drafts of this book. The generous work of academic experts is essential for establishing the authority and quality of our publications. We acknowledge with gratitude the contributions of these otherwise uncredited readers.

This book was set in Stone Serif and Stone Sans by Westchester Publishing Services. Printed and bound in the United States of America.

Library of Congress Cataloging-in-Publication Data

Names: Gärdenfors, Peter author | Osta-Velez, Matias author
Title: Reasoning with concepts : conceptual spaces as a framework / Peter Gärdenfors and Matias Osta-Velez.
Description: Cambridge, Massachusetts : The MIT Press, [2026] | Includes bibliographical references and index.
Identifiers: LCCN 2025037120 (print) | LCCN 2025037121 (ebook) | ISBN 9780262053402 paperback | ISBN 9780262053419 pdf | ISBN 9780262053426 epub
Subjects: LCSH: Reasoning (Psychology) | Cognition | Concepts
Classification: LCC BF442 .G37 2026 (print) | LCC BF442 (ebook)
LC record available at https://lccn.loc.gov/2025037120
LC ebook record available at https://lccn.loc.gov/2025037121

EU Authorised Representative: Easy Access System Europe, Mustamäe tee 50, 10621 Tallinn, Estonia | Email: gpsr.requests@easproject.com

Contents

Preface

This is the third book about the theory of conceptual spaces. The first, *Conceptual Spaces: The Geometry of Thought* (MIT Press 2000), presented the framework and applied it to some philosophical problems. The second, *Geometry of Meaning: Semantics Based on Conceptual Spaces* (MIT Press 2014), applied the theory to problems in the semantics of natural language.

The present book applies conceptual spaces to problems in cognitive psychology, further extending the theory and its scope. This time Peter Gärdenfors has joined efforts with Matías Osta-Vélez, a philosopher working at the intersection of philosophy of science and cognitive science. Matías came to visit Peter in Lund in 2019, and he proposed several areas of reasoning with concepts where he thought that conceptual spaces would be a useful modeling tool. So began a very fruitful collaboration that has resulted in a series of articles and ultimately in writing this book.

A central motivation for this book is the realization that in philosophy and psychology, theories of concepts and theories of reasoning tend to develop independently. We aim to show that this divide is unfounded and that both areas can mutually inform each other given the right theoretical framework.

We have not done our work in isolation but have been helped by several people. We wish to thank Andreas Stephens for permission to use material from the article Gärdenfors and Stephens (2018) that forms the basis for parts of chapter 4. We also thank Igor Douven, Shira Elqayam, Patricia Mirabile, and Steven Verheyen, who have joined us in testing some of the predictions of our models (Douven et al. 2022, 2023, 2025). Many other friends and colleagues have given feedback to the ideas on this book throughout its development. Special thanks are due to Matteo de Benedetto,

Peter Brössel, Corina Strößner, Gerhard Schurz, Guillermo Nigro, Alejandro Chmiel, Annika Schuster, Mariela Aguilera, Max Kistler, Stephan Hartmann, Philipp Koralus, James Hampton, Paul Egré, Sebastian Scholz, Henri Stephanou, Horacio Lena, Michal Hladky, Washington Morales, Álvaro Peláez, Martin Jönsson, and Nina Poth. Matías is especially grateful to Lena Peichl for her constant support and patience throughout this project. The editors of the MIT Press have done an excellent job in turning our manuscript into a book.

1 Introduction

As the story goes, reasoning is the pinnacle of human cognition, underpinning everything from everyday decision-making and planning to the greatest achievements across the spectrum of human intellectual disciplines. We are often taught to see reasoning as a fundamentally logical process, governed by a finite set of universal, abstract rules. Yet, when we observe commonsense reasoning in action, we find many features that challenge this view and resist purely logical explanation.

In this book, we argue that much of our reasoning depends not on formal rules operating at the level of sentences or propositions but on the *structure of concepts*—that is, the ways in which semantic information about categories is organized and represented in the mind. Imagine, for instance, that you learn that dogs can suffer from parvovirus. What other conclusions might you draw? It seems reasonable to infer that wolves could also suffer from this virus, given their similarity to dogs. However, you would likely hesitate to extend this inference to rabbits or turtles, despite all being animals. This kind of reasoning, which psychologists call "category-based induction," is not grounded in formal properties of propositions but in semantic relations between concepts.

Similarly, our use of generic statements—such as "Birds fly" or "Sharks are dangerous"—in everyday reasoning shows that conceptual structure often takes precedence over logical structure. For instance, we accept without hesitation that "Birds fly" is true even though we also believe statements that contradict it, like "Penguins do not fly." Even with these tensions, generics still play a very important role in our cognitive lives.

Another example can be found in our use of analogies in natural language. Consider why "The foot is to the leg as the hand is to the arm" feels intuitively right, while "The foot is to the leg as the liver is to the stomach" strikes us as wrong. This contrast does not stem from differences at the level of *logical form*, but rather from the specific mereological (part-whole) relationships encoded in our concepts.

Psychologists have studied these and other forms of concept-based reasoning from multiple theoretical and methodological perspectives. This book develops a unified account of these processes drawing on the theory of conceptual spaces (Gärdenfors 2000, 2014). Within this framework, we define measures of *similarity*, *typicality*, *diagnosticity*, and semantic *coherence*, that will enable us to explain these inferential mechanisms and others, while generating new testable predictions. In the remainder of this introductory chapter, we discuss the influential view in philosophy and cognitive psychology that all reasoning must be based on logical or probabilistic aspects of propositions, and we argue that the structure of lexical concepts plays a central role that must be integrated into any comprehensive theory of reasoning. Finally, we introduce explicit definitions of the measures, which will serve as the basis for our models.

1.1 What Is Reasoning?

Reasoning can be loosely defined as the process of drawing conclusions from information available to the cognitive agent. Philosophers and cognitive scientists distinguish between two fundamental types: *ampliative* or *conservative* reasoning (Hintikka 1998; Osherson et al. 1993). For example, if one infers that it just rained from seeing a wet sidewalk, then one is going beyond the initial information, and therefore one's reasoning is ampliative. If one infers that 28 is divisible by 2 because it is an even number and even numbers are divisible by 2, then this is a conservative inference as the information in the conclusion is already contained in the premises.

Crucially, reasoning is normatively constrained. When we reason, we do not jump arbitrarily from one thought to another, but our inferences are based on standards of relevance and justification. For instance, in concluding that it rained from seeing the wet sidewalk, our reasoning was guided by the belief that "wetness" is evidentially relevant to "rain," which is supported by our background knowledge of how rain typically causes wet surfaces. This characteristic of reasoning sets it apart from other mental processes such as free association or daydreaming, where the transitions between thoughts are rather idiosyncratic, without any concern for justification or evidence (see Buckner 2019).[1] The challenge for a theory of reasoning is, then, to specify the mechanisms underlying rational inference that demarcate it from other mental processes.

1.2 Form and Content in Reasoning

How can we distinguish reasoning from other kinds of mental acts? Early research on the psychology of reasoning saw classical logic as the key to answering this question. Influenced by early analytic philosophy, researchers hypothesized that human reasoning was a logic-like system equipped with a finite set of inference rules operating at the level of propositions (Oaksford and Chater 1991; Over and Evans 2024). Central to this paradigm is a sharp distinction between *form* and *content*: Inference is thought to depend on the *structural* or *syntactic* features of propositions, irrespective of their specific subject matter. In other words, we reason by manipulating propositional representations according to their logical properties and independently of their semantic or thematic content.

The form-content distinction has its roots in Aristotle's analysis of argumentation (Aristotle 1965, 29). Aristotle famously noticed that, despite their varied subject matters, arguments followed recurring structural patterns, which he captured by introducing *syllogisms*, generalized argument structures represented by schematic letters (see Corcoran [2006] for a detailed explanation). Crucially, Aristotle posited that the validity of an argument was based solely on its formal properties, as reflected in syllogisms, and was unaffected by its "material" properties—that is, the specific content or subject matter. This led to the fundamental insight that logic must operate independently of the content of the concepts involved in arguments.

While Aristotle used schematic letters to capture argument patterns, modern logic takes this abstraction further by identifying a group of words in natural language that tie together all logical reasoning. These terms, known as *logical constants* and including words such as "and," "or," "if . . . then," and "not," are what Ryle called "topic-neutral" (1945, 116): Rather than referring to things in the world, they derive their meaning solely from their structural role in connecting concepts and propositions. A crucial feature of logical constants is that they are *truth-functional*: They act like operators that determine whether complex statements are true or false based solely on the truth values of their components. These two key features, *topic-neutrality* and *truth-functionality*, allowed logicians to define inferential validity as a function of logical form alone, independent of any specific content (see Read 1994). This approach consolidated the idea that the normative structure of reasoning is fundamentally a matter of form rather than content and gave rise to what we call the "formalist view of reasoning," whose central claim is that if human reasoning is indeed logical, it must be governed by formal-syntactic rules.[2,3]

One of the most influential defenses of the formalist view of thinking was articulated by Fodor (Fodor 1975, 1980; Fodor and Pylyshyn 2015). Fodor argued that the central challenge of cognitive psychology is to provide a causal explanation of an inherently intentional phenomenon such as thinking. This challenge requires explaining how thoughts maintain their intentional and semantic properties across coherent inferential transitions. For Fodor, this meant that any plausible explanation of cognition had to show how reasoning can be computationally reconstructed. He argued that this could only be achieved by conceptualizing the mind as a syntax-driven machine performing formal operations in a "language of thought" with both syntactic and semantic properties. Inspired by classical proof theory, this approach postulates that syntactic properties enable causal transitions between thoughts while mirroring their semantic properties, making rational thought possible (Fodor 1985, 93, 1987, 19).

The psychology of reasoning emerged within the computationalist paradigm, significantly influenced by developments in mathematical logic during the first half of the twentieth century. This new field adopted a predominantly logicist orientation. Piaget's work regarded logic as a mathematical model for high-level cognition at large (Winstanley 2022), while subsequent theories directly proposed logic as both a computational and an algorithmic model of reasoning. Examples include the mental logic theory (Braine 1994) and Rips's (1994) deduction-system hypothesis, where all inferential transitions depend on the application of domain-general rules of inference.[4]

1.2.1 Psychological Evidence Against the Formalist View of Reasoning

A crisis broke out for the formalist view of reasoning after empirical studies began to reveal that people too often fail to follow logical rules (Perkins 2002). In particular, the Wason selection task (Wason 1968) proved that people struggle with deductive tasks involving simple inferential rules such as *modus ponens* and *modus tollens*. In this task, participants see four cards with inscriptions A, D, 2, and 7 and must verify the rule: "If a card has a vowel on one side, then there is an even number on the other." The correct solution requires flipping the cards A and 7, but most people choose A alone or A and 2. Interestingly, performance improves when the task is designed with materials that are less abstract and more familiar to people. Wason and Shapiro (1971) created a thematic version using the rule "Every time I go to Manchester, I travel by car." This version saw 62 percent correct responses, compared to 12 percent for the abstract version. According to Wason and Shapiro, the improved performance suggests that reasoning is driven more by content-related mechanisms than syntactic ones.

This finding led to a surge of interest in "content effects" within the psychology of reasoning (Manktelow and Evans 1979; Griggs and Cox 1982). In particular, prior knowledge and experience seem to play a significant role in deductive reasoning (for a review, see Evans and Feeney 2004). Researchers concluded that people do not typically follow domain-general rules of inference in problem-solving contexts but rather seek information and counterexamples from domain-specific memories.

The findings align with another phenomenon commonly observed in reasoning: the *belief bias*—a tendency to evaluate arguments based on the believability of their conclusions rather than their logical structure. For instance, people tend to reject the argument "No addictive things are inexpensive. Some cigarettes are inexpensive. Therefore, some cigarettes are not addictive," even though it is logically valid. Conversely, people are more likely to accept as valid the argument "No addictive things are inexpensive. Some cigarettes are expensive. Therefore, some cigarettes are addictive," even though it is logically invalid (Evans et al. 1983; Klauer et al. 2000). This shows that there is a complex interaction between *validity* and *believability* in argument evaluation, suggesting a strong inclination to judge arguments based on the plausibility of the conclusion rather than its logical structure.

Empirical evidence consistently demonstrates that everyday reasoning focuses more on the *meaning* of concepts and the content of background beliefs than on logical structure. This suggests a strong interaction between inferential mechanisms and the content of the extra-logical terms, indicating that—contrary to the formalist view—reasoning cannot be exclusively logic-based. The explanatory limitations of formalist approaches in accounting for these findings have led to their decline within the psychology of reasoning (Evans 1991; Oaksford and Chater 1995; Johnson-Laird 2010).

1.3 The Bayesian Alternative

An influential alternative to the formalist view of reasoning comes from Bayesian cognitive science (Oaksford and Chater 2007; Tenenbaum and Griffiths 2001). Rather than focusing on logical forms and truth-preserving inference, Bayesian models characterize reasoning as the process of updating *degrees of belief* in light of new evidence. Rational inference, in this view, is not about establishing conclusions with absolute certainty but about dynamically adjusting the probabilities of beliefs based on new information or hypotheses.

The Bayesian approach is also "propositionalist" in that it treats reasoning as belief updating over probability distributions assigned to propositions. However, it diverges significantly from logicist approach by interpreting conditional statements of the form "*If p, then q*" not as material conditionals but as statements about conditional probabilities, specifically $P(q|p)$. This shift allows Bayesian models to account for the nonmonotonic nature of human reasoning, where conclusions are not strictly valid or invalid but more or less likely given prior knowledge and new evidence (we will discuss this in depth in chapter 5). Crucially, Bayesian theorists do not disregard the role of content in reasoning: Background knowledge influences prior probabilities, shaping how people update their beliefs and reach conclusions. Formally, prior estimates encode how strongly we believe a given proposition before encountering new evidence, and Bayes' rule prescribes how those beliefs should be revised accordingly.

While content is included through prior probabilities, a key question remains: Where do these priors come from in the first place? Proponents of Bayesian cognitive science often argue that priors reflect accumulated knowledge—innate constraints, learned associations, or conceptual commitments. However, Bayesian theory alone does not specify how this conceptual content is represented or how it is translated into probability distributions. This gap highlights a central challenge for any account of cognition that relies on probabilities: We need to explain not only how semantic structure, similarity relations, and contextual knowledge are encoded in priors and probabilities but also how people can feasibly handle the computational demands of such probabilistic inferences.[5]

In what follows, we present a model that challenges the propositionalist and formalist approach by focusing on lexical concepts instead of propositions. While not meant to address the limitations of Bayesian reasoning, this model offers an alternative perspective in which conceptual knowledge is not merely an input to inference but the very foundation that guides and constrains it.[6]

1.4 Concepts and Reasoning

This book aims to bridge the gap between theories of reasoning and theories of concepts, which have long been treated as separated areas in psychology and philosophy. Psychologists have traditionally viewed the mind as comprising independent, domain-general faculties such as memory, categorization, perception, and reasoning, which they believed could be studied separately (see Fodor 1983). Philosophers, on the other hand, have focused on normative frameworks for understanding reasoning, relying heavily on a formalist view that emphasizes logical or probabilistic structure over content.

However, there seems to be no principled reason to keep the study of concepts separate from the study of reasoning. *Concept* is a key notion in theories of semantic memory and knowledge representation, and if reasoning draws on knowledge, we must consider the connection between inference and conceptual structure. Indeed, it is difficult to define "concept" without implications for inference and vice versa. For example, possessing the concept "grandfather" requires being able to make inferences about related concepts such as "son," "male," "father," and so on (Brandom 2001). Conversely, *inference* is an informational mechanism and, as such, is regulated by multiple semantic/conceptual constraints (Keil 1979; Johnson-Laird and Byrne 1991; Schurz 1991).

Like most cognitive scientists and philosophers, we assume concepts underlie most high-level cognitive processes, particularly categorization and reasoning (Rosch 1988; Murphy 2002; Machery 2009). A central tenet of our theoretical framework is (contra Fodor) that concepts have complex structures with various components and features, which are relevant in different ways to the cognitive tasks for which they are recruited. In particular, in this book we will demonstrate how different aspects of the structure of concepts and conceptual systems can be used to explain various inferential processes.

Following Evans (1989), we argue that theories of reasoning cannot be functionally independent of theories of concepts. No computational or algorithmic description of reasoning would be complete without referencing the organization of knowledge in semantic memory and how this organization influences inferential mechanisms. Just as the study of deductive reasoning requires taking into account multiple logical aspects of language, the study of ampliative reasoning needs to consider how background knowledge and conceptual content shape inference. Cognitive psychology has identified several types of inference patterns clearly rooted in our conceptual knowledge, including lexical inference, induction (including category-based induction), nonmonotonic reasoning with generic sentences, and analogy. While these inferences have been explained using different theoretical frameworks, we aim to present a unified model based on conceptual spaces.

An important theoretical premise of this book, and one that diverges from the dominant tradition, is that the kind of knowledge used in reasoning with concepts is different from propositional knowledge. Ryle (1949) introduced a classical dichotomy between "knowledge-that" and "knowledge-how," where knowledge-that is knowledge about propositions or facts and knowledge-how concerns the skill to perform some practical task. We argue, however, in chapter 4 that there is a third kind of knowledge that we call "knowledge-what," a form of conceptual knowledge that goes beyond Ryle's (1949) classical dichotomy. Knowledge-what encompasses the ability to categorize and comprehend the semantic relationships between categories and their properties (Gärdenfors and Stephens 2018). Accepting this form of knowledge, as we will show, is essential to understand both the problem of induction and concept-based inferences more broadly.

On the formalist view of reasoning, the fundamental units of knowledge-that are propositions (expressed in sentences) and are typically modeled through logical or probabilistic frameworks. When it comes to knowledge-what, however, we propose that conceptual spaces provide a more appropriate framework for describing the structure of concepts and how they are used in reasoning. The theory of conceptual space will be presented in the following chapter.

1.5 Similarity, Typicality, Diagnosticity, and Coherence

The models we propose for explaining concept-based reasoning focuses on four explanatory variables: *similarity*, *typicality*, *diagnosticity*, and *coherence*. Although these notions are used in categorization research, there are no specific measures of each of them within a single formal framework. A key contribution of this book is to provide such measures, using distances in conceptual spaces. We take similarity, together with the notion of prototype, as starting points and use them to define typicality, diagnosticity, and coherence. In what follows, we describe these notions informally and return to a more detailed and formal treatment in section 2.7.

1.5.1 Similarity

The notion of similarity is arguably the cornerstone of any theory of concepts and has played an important role in the psychology of inductive reasoning. The development of categorization is essentially the refinement of our ability to differentially respond to stimuli (Shepard 1987). Analogously, concept formation is essentially the construction of equivalence classes of similar stimuli. Quine summarized the role as follows: Similarity "is the basis of all learning, all habit formation, all expectation by induction from past experience; for we are innately disposed to expect similar events to have sequels that are similar to each other" (Quine 1995, 253).

The exact meaning of similarity has been hotly debated. Goodman (1972, 437) famously challenged the very meaningfulness of similarity, calling it "insidious" and dismissing it as "a pretender, an impostor, a quack." On the other hand, he also states, "First, we must recognize that similarity is relative and variable, as undependable as indispensable. Clear enough when closely confined by context and circumstances in ordinary discourse, it is hopelessly ambiguous when torn loose" (1972, 444). The main point of Goodman's argument is that there is nothing like overall similarity that can be universally measured, but we always have to say in what respects two things are similar. Similarity judgments will thus become crucially dependent on the *context* in which they occur.

In this book, following Shepard (1987), we describe similarity to be inversely related to distances in a conceptual space. In other words, the closer the points representing two objects are in a conceptual space, the more similar they are judged to be. This makes our notion of similarity different from that of Tversky (1977), which is based on comparing the number of properties that two objects have in common with the properties where they differ. We will return to Tversky's criticism of similarity as a distance measure in section 2.6.1. There, we will also show how the context dependence of similarity can be modeled in terms of the *salience* of the dimensions in a conceptual space.

1.5.2 Typicality

Thanks to the work of Rosch (1975) and her followers, we know that concepts have a prototypical structure in which some exemplars are more representative of the category than others (e.g., within the concept BIRD, a robin is considered more typical than a penguin). This feature of semantic representation is particularly important for cognitive tasks such as categorization, in which the ease and speed of classification varies accordingly to the typicality of the instance being classified. Prototype theory allows for representing graded membership and degrees of typicality (Rosch et al. 1976; Hampton 2007); that is, we can represent objects in the space as being more or less typical instances of their categories according to their position relative to a prototype.

However, what is needed in many psychological contexts is not just to judge how typical different objects are but to determine how typical a particular *property* is for a certain concept (e.g., *having feathers* is a more typical property of BIRD than *singing*). Such a measure, it will turn out, is central for assessing the validity of generics, modeling category-based induction, and validating analogies—in other words, for the main types of reasoning with concepts studied in this book. Section 2.6.2 proposes a formal measure of the typicality of a property that will be extensively applied throughout the following chapters.

1.5.3 Diagnosticity

Diagnosticity is, in a way, a more subtle notion than the two previous ones. It concerns how effectively properties distinguish a concept from related concepts. In brief, a property is diagnostic when it maximizes the dissimilarity between a concept and the concepts in its contrast class, speeding up categorization (Tversky 1977; Chin-Parker and Ross 2004). For example, the property *having a mane* is diagnostic of the concept LION, since having manes (at least in adult males) sets them apart from other felines. Analogously, *having scales* is diagnostic of FISH with respect to the standard partitioning of the concept ANIMAL.

Diagnostic properties are particularly important in reasoning because they carry the most informational content about a concept relative to similar concepts. This explains why they play a crucial role in concept learning and why generic sentences often feature diagnostic properties, as we will show in chapter 6.

1.5.4 Coherence

Coherence is what we could call a *global* property of concepts (it concerns the concept as a whole rather than any individual feature or exemplar). Specifically, coherence refers to the extent to which the features associated with a particular concept are perceived as "belonging together" or, in other words, to the perceived unity among exemplars based on shared features, functions, or other underlying structures (Smith and Medin 1981). For example, the concept of SPORT CAR is more coherent than the concept of FURNITURE: Most exemplars of sports cars look very similar and have similar dispositional properties. In contrast, the furniture concept encompasses a far more diverse range of objects with varying forms and functions.

There is substantial evidence that differences in the coherence of our concepts impact their role in induction and categorization. However, a limitation of this notion is that it is particularly difficult to formalize. In chapter 9, we will provide a way to model conceptual coherence and discuss its relation to inductive reasoning.

1.6 Main Themes of the Book

As a conclusion to this introductory chapter, we present a summary of the main contributions of this book. We develop formal models of many psychological phenomena related to different forms of reasoning with concepts—particularly lexical entailment (chapter 3), induction (chapters 4 and 7), reasoning with expectation (chapter 5), generics (chapter 6), and analogies (chapter 8). While the cognitive psychology literature has produced several models of these processes, they have typically been treated as disconnected phenomena. In contrast, the framework of conceptual spaces (chapter 2) enables us to unify these diverse forms of reasoning. This unification is accomplished through measures of similarity, typicality, and diagnosticity, which we define in terms of distances and prototypes within conceptual spaces. Although we do not provide computational implementations of these models, we outline how they might serve as foundations for implementing concept-based reasoning in artificial systems.

Beyond these forms of reasoning, we also advance new theoretical accounts of conceptual coherence (chapter 9) and essentialism about concepts (chapter 10), further extending the explanatory power of our approach.

2 Conceptual Spaces as a Framework

2.1 Three Levels of Concept Representations: Symbolic, Conceptual, and Subconceptual

Within the cognitive sciences, there have been two dominating approaches to the problem of representing knowledge. The first approach, known as the *symbolic* approach, is based on the idea that cognitive systems should be modeled by Turing machines. This is what we described as the formalist approach in the previous chapter. The second approach is *connectionism*, which models cognitive systems by artificial neural networks. Both approaches have their advantages and disadvantages. They are often presented as competing paradigms, but since they address cognitive problems at different levels of granularity, they should be seen as complementary methodologies (Fodor and Pylyshyn 1988; Dinsmore 2014).

There are, however, aspects of cognitive phenomena for which neither symbolism nor connectionism seem to offer appropriate modeling tools. In this book, we build on a third form of representing information that builds on *geometric* structures rather than symbols or connections between neurons. We call it *conceptual* representations since we think that the essential aspects of concept formation and reasoning with concepts should be described on this level. It is *geometric* in the sense that notions such as *dimension*, *betweenness*, *convexity*, and *distances* will be central in the framework. Again, conceptual representations should not be seen as competing with symbolic or connectionist representations. Rather, the three kinds can be seen as three *levels* of representations of cognition (Gärdenfors 2000).

2.2 Dimensions and Domains

Conceptual spaces have been developed as a research program in semantics studying the structure of concepts and their interrelations (Gärdenfors 2000, 2014). The approach builds on two main ideas: (1) concepts are composed of clusters of quality dimensions, many of which are generated by sensory inputs such as color, size, and temperature; and (2) concepts have a geometric or topological structure that results from integrating of the structures of the individual dimensions.

Quality dimensions can be *integral* or *separable*. They are integral when it is not possible to assign a value to an object on one dimension without simultaneously assigning a value on another dimension (Maddox 1992; Melara 1992). For instance, it is not possible to attribute a value to the pitch of a tone without attributing one to its loudness, or determine the hue of an object without also determining its brightness. When quality dimensions are not integral, they are called *separable*.[1] Several psychological tests exist for determining whether dimensions are integral or separable (Garner 1974; Maddox 1992; Melara 1992).

In children's semantic development, some domains are gradually separated and can be attended to as separable set of dimensions. For example, two-year-olds can represent whole objects, but they cannot reason about the dimensions of those objects. Goldstone and Barsalou (1998, 252) note:

> Evidence suggests that dimensions that are easily separated by adults, such as the brightness and size of a square, are treated as fused together for children . . . For example, children have difficulty identifying whether two objects differ on their brightness or size even though they can easily see that they differ in some way. Both differentiation and dimensionalization occur throughout one's lifetime.

A classic example from psychology is Piaget's (1972) studies of "conservation." Children under the age of five cannot separate the volume of liquids from their height. When choosing between two glasses of lemonade, they pick the glass with the highest level of lemonade even though that glass is very narrow and the other is wide. If the lemonade is poured into a shallow container, they believe there is less of it. Only later do they learn that the volume of a liquid is conserved between containers and not always correlated with height. In other words, they have learned to separate the dimension of height from that of volume. In this way, their conceptual space is expanded with a new domain.

We define the central notion of a *domain* as a set of integral dimensions that are separable from all other dimensions. For instance, human color properties are composed of three fundamental parameters of color perception: hue, saturation, and brightness. Hue can be represented as a circle (as in the traditional color circle), saturation corresponds to an interval of the real line, and brightness (which varies from white to black) is a linear dimension with endpoints (see figure 2.1).[2] Any color perception can be mapped onto specific values of hue, saturation and brightness. More generally, different colors can be described as *convex regions* of possible values of these three parameters. A region is convex when, for

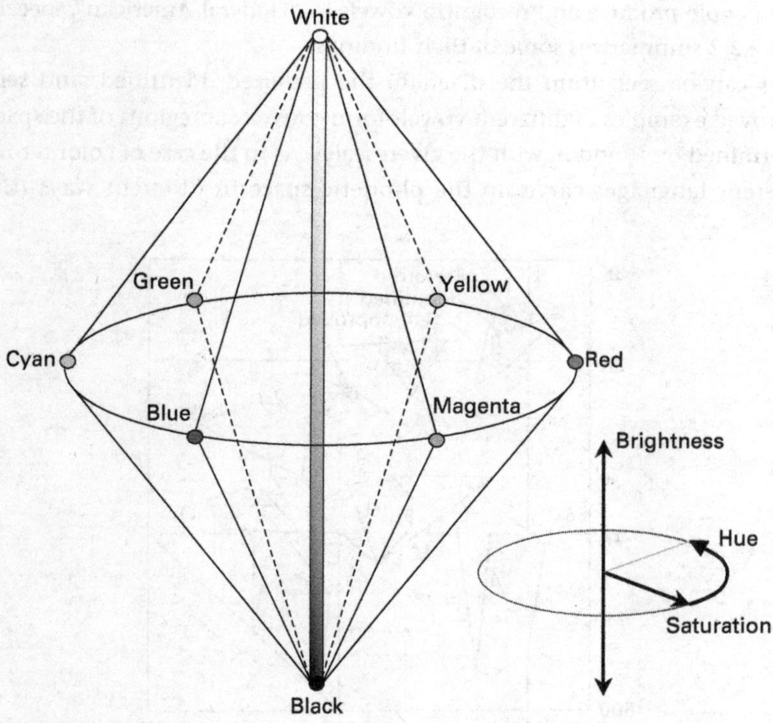

Figure 2.1
The Natural Color System (NCS) (hue, saturation, brightness) color space is visualized as a double cone. Hue (0°–360°) is represented by the angular dimension, saturation (0–1) by the radial distance from the central axis, and brightness (0–1) by the vertical axis. This model separates color information (hue, saturation) from brightness, closely aligning with human color perception.

every pair of points *x* and *y* in the region, all points *between* them are also in the region.

Another illustration of how the convexity of regions determines concepts and categorizations is the phonetic identification of *vowels* in various languages. According to phonetic theory, what determines the quality of a vowel are the relations between the basic frequency F_0 of the sound and its formants (higher frequencies that are present at the same time). In general, the first two formants F_1 and F_2 are sufficient to identify a vowel. This means that the coordinates of two-dimensional space spanned by F_1 and F_2 (in relation to a fixed fundamental frequency F_0) can be used as a fairly accurate description of a vowel. Fairbanks and Grubb (1961) investigated how people produce and recognize vowels in "General American" speech. Figure 2.2 summarizes some of their findings.

As can be seen from the diagram, the preferred, identified, and self-approved examples of different vowels form convex subregions of the space determined by F_1 and F_2 with the given scales. As in the case of color terms, different languages carve up the phonetic space in different ways (the

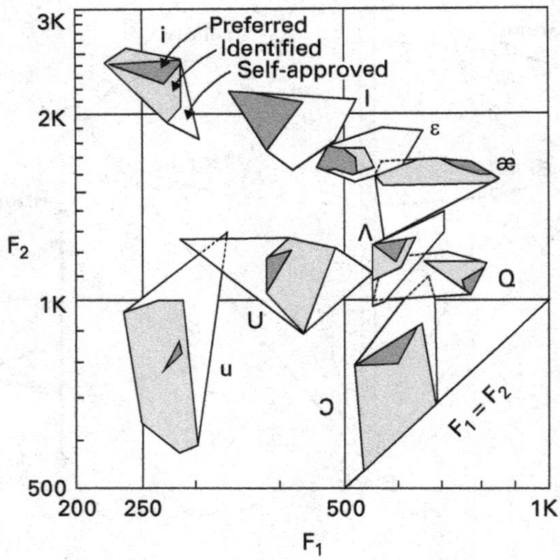

Figure 2.2
The vowel space of American English (from Fairbanks and Grubb 1961). The scale of the abscissa and ordinate are the logarithm of the frequencies of F_1 and F_2. Reprinted with permission from the American Speech-Language-Hearing Association.

number of vowels identified in different languages varies considerably), but each vowel in a language corresponds to a convex region of the formant space.

A domain that can be studied both from a psychological and a neuroscientific perspective is the *emotion space*. Many human emotions are shared with mammals, birds, and other species. All these species show clear indications of fear, anger, and pleasure. These emotions and their expressions are partly regulated by evolutionarily old brain structures (LeDoux 2003). Several models to represent the psychological space of human emotions have been proposed, many of which are extensions of Russell's (2003) two-dimensional space. Russell aims to identify core affects, which he defines as "a neurophysiological state that is consciously accessible as a simple, nonreflective feeling" (Russell 2003, 147). On his analysis, core affects involve "an integral blend of hedonic (pleasure–displeasure) and arousal (sleepy–activated) values" (ibid.). These two dimensions form his emotional space.

Russell shows that the meaning of most emotion words can be mapped onto a circumplex defined by these two dimensions (figure 2.3). There are other models of the psychological emotion space that include a third dimension—for example, a "dominance" dimension that expresses the controlling nature of the emotion (Mehrabian 1996). For example, while both fear and anger are unpleasant, anger is dominant, while fear is nondominant.

A central question for all forms of knowledge representation is how the representations are grounded outside the mind (Harnad 1990; Taddeo and Floridi 2005). Within the cognitive sciences, the grounding can take place in the *body* (embodied cognition) and in the *external world* (situated cognition). Within cognitive linguistics, these forms of grounding have become central parts of the semantics of words (Johnson 1987; Lakoff 1987). As can be seen from the examples of domains presented in this section, the most fundamental domains are sensory and emotional and thus grounded in the body or the external world. There exist, however, more abstract domains—for example, *time* and *kinship*—that are not directly grounded. Furthermore, in the conceptual spaces underlying scientific theories (Gärdenfors and Zenker 2010, 2013), the grounding of the observable dimensions takes place via the measurement procedures, while the theoretical variables (e.g., force in Newtonian mechanics or energy in relativity theory) are not grounded.[3]

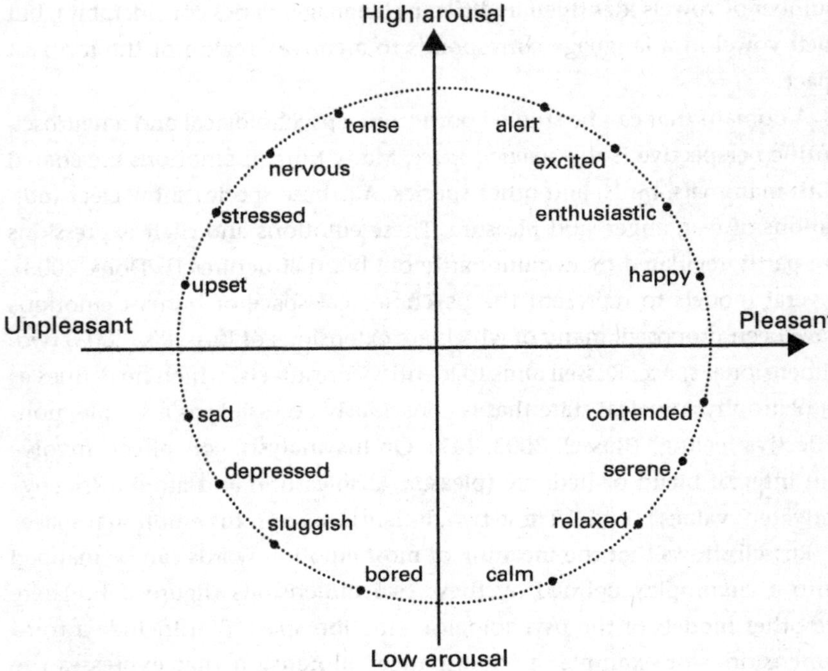

Figure 2.3
Russell's (1980) two-dimensional emotion space has two basic dimensions of valency and arousal and different emotions arranged in a circular structure.

In brief, the theory of conceptual space can accommodate the role of the body and the external world in its representations but is not limited to these structures.

2.3 From Similarity to Dimensions

The dimensions of a conceptual space cannot be derived immediately from the perceptions or actions of the subjects but must be inferred by other means. In research on concept formation, it is often assumed that the *similarity* between two stimuli mirrors the distances between their representations in the underlying psychological space. A common assumption in the psychological literature (Shepard 1987; Nosofsky 1988, 1992; Hahn and Chater 1998) is that similarity is an exponentially decaying function of distance. If *sim* (*i*, *j*) expresses the similarity between two objects *i* and *j* and $d(i, j)$ their distance, then the following formula, where c is a general "sensitivity" parameter, expresses the relation between the two measures:

$$sim(i, j) = e^{-c\, d(i, j)}. \tag{2.1}$$

Shepard (1987) calls this the universal law of generalization, arguing that it captures the similarity-based generalization performances of subjects in various settings. An underlying motivation for the equation is that matching and mismatching properties are combined multiplicatively rather than additively (Nosofsky 1992; Medin et al. 1993, 258). Given some additional mathematical assumptions, this corresponds to an exponential decay of similarity.

There are several statistical techniques for identifying the dimensions of a conceptual space, given a dataset of similarity judgments. Here, we briefly introduce two of the most well-known methods, namely *multidimensional scaling* (MDS) and *principal component analysis* (PCA).

Multidimensional scaling is used to determine the number of dimensions in the underlying phenomenal space and the scaling of the space (Kruskal 1964). Starting from subjects' judgments about the similarities of a class of stimuli, the goal is to obtain as high a correlation as possible between the similarity judgments of the subjects and the corresponding distances in the estimated dimensional space. Similarity judgments can be numerical, but they are often given in an ordinal form obtained from a scale ranging from "very similar" to "very dissimilar." The judgments given by the individual subjects are normally averaged before they are fed into an MDS algorithm.

The investigator chooses the number n of dimensions in the space to be estimated and the metric (normally Euclidean or city-block) to be used in defining distances in the space. Starting from an initial assignment of coordinates to the stimuli in an n-dimensional space, the MDS algorithm then systematically adjusts the coordinates to achieve a progressively better fit to the data from the similarity judgments. The degree of misfit between the data and the estimated space is normally measured by a "stress function."[4] The algorithm stops when the stress of the estimated space no longer decreases. The higher the n of the estimated space, the smaller the resulting minimal stress. Thus, an arbitrarily good fit can be achieved by increasing n. A methodological problem is, therefore, how to decide what value of n to use in a model. Unless there are strong a priori reasons for assuming that the underlying phenomenal space has a certain number of dimensions, one often looks for a (rather small) number n where the computed stress of $n+1$ dimensions is not significantly smaller than the stress of n dimensions. One problem with MDS is that it can be difficult to give a psychological interpretation of the dimensions generated by the algorithm (given a choice of n).

Principal component analysis exploits redundancies in input data (Schyns 1991, 471–472; Schyns et al. 1998, 15). If the dimensions of the input data are correlated, principal component analysis finds a number (determined by the user) of orthogonal directions in the dimensions of the input data that have the highest variation. Thus, the first principal component is the spatial direction in the dataset with the highest variation, and which thus is the maximally "explanatory" dimension. However, the process is not without drawbacks. Schyns et al. (1998, 15) have the following comment: "In general, however, the featural interpretation of principal components is often difficult because orthogonal directions of highest variance have little connection to the best projections for categorization. That is, there are no *psychological* constraints on the principal components."

2.4 A Classification of Concepts

In this book, we use the term "concept" to refer to mental representations and the term "category" to refer to their extensions (be it a *kind* or a *class*). This means that we adopt a cognitive approach to concepts. In contrast, many philosophers attempt to develop realist theories, in which concepts are assumed to exist independently of human (or animal) minds (Katz 1990; Peacocke 1992).[5] As the metaphysical problems of the theory of concepts extend beyond the scope of this book, we will not address them here. However, for our purposes, we advance a minimal classification of concepts based on the types of extensions they are intended to capture.

Many concepts in natural language, particularly those expressed by nouns, cannot be defined within a single domain but rather as clusters of properties. This leads us to distinguish between *properties* and *object concepts*. Object concepts are representations (enduring in memory or not) of collections of entities denoted by a noun or a phrasal nominal (e.g., "green thing," "old lawyer," "things to bring to the beach"). Throughout this book, we will use a specific notation to differentiate between these notions.

Object concepts will be denoted using small caps (e.g., DOG or CAR), while properties will appear in italics (e.g., *brown*, *warm*).

Apart from properties and object concepts, there also exist relational and abstract concepts. However, we will not attempt a further classification of these. Many abstract concepts, such as "freedom," "justice," and "inflation," are difficult to analyze in terms of domains (Borghi et al. 2017). Our focus in this book will primarily be on properties and object concepts, with occasional incursions into relational concepts and concepts representing actions. This choice is motivated by our main areas of interest: induction, generics, and analogies. In these types of reasoning, the analysis primarily concerns relations between object concepts and properties.

At a second level of classification, we divide object concepts into *kind concepts* and *class concepts*. Kind concepts refer to collections of entities sharing multiple properties and are linguistically expressed by lexical nominals (e.g., "dog," "apple," or "hammer").[6] They are stable representations of more or less coherent clusters of correlated properties that align with people's notion of naturalness (Prasada et al. 2013; Strößner 2022).

Class concepts refer to any object concept that is not a kind concept. These encompass exemplars that do not necessarily share more than one property and, consequently, are not understood as genuinely similar or being part of the same kind. Class concepts include *ad hoc concepts*, such as "things to bring to a picnic" (Barsalou 1983), and concepts defined by adjective-noun combinations, like "white cups." Our terminology is summed up in table 2.1

There is extensive evidence that properties, class concepts, and kind concepts have different cognitive profiles, especially concerning learning and

Table 2.1
Our classification of concepts

	CONCEPTS		
Properties	Object concepts		Other concepts (e.g., abstract or relational)
White, small, warm, round	**Kind concepts** DOG, TIGER, HAMMER, ROUND	**Class concepts** LARGE WHITE THINGS, THINGS TO BRING TO A PICNIC	DEMOCRACY, HEIR, JUSTICE, INFLATION
CATEGORIES	Classes of things		

processing (Rosch et al. 1976; Barsalou 1983; Markman 1989; Smith et al. 1992; Gasser and Smith 1998; Prasada et al. 2013). In this book, however, we will focus on *structural* aspects of their representation and their relation to inference (Medin et al. 2000, 123).

Regarding the distinction between kind concepts and class concepts, we adopt the "distinct representation hypothesis" (Prasada et al. 2013), which posits that these two types function differently in reasoning. Prasada et al. (2013) present a series of experiments supporting this hypothesis, and we offer several kinds of further support for it throughout this book.

In considering the nature of properties based on similarity spaces, the following general constraint has been proposed (Gärdenfors 2000, 71):

> Criterion *P*: A *natural property* is a convex region of a single domain.[7]

Convexity is a principle of cognitive economy; handling convex sets puts less strain on learning, your memory, and your processing capacities than working with arbitrarily shaped regions (Gärdenfors 2024b, 2024c).

Kind concepts are not just bundles of properties. For most such concepts, the domains that compose them can be correlated in different ways. For example, in the case of the concept FRUIT, properties like *size* and *weight*, or *ripeness*, *color*, and *taste* covary. These covariations generate *expectations* crucial for inferential procedures that exploit semantic properties.[8] For this reason, the representation of a kind concept in a conceptual space also includes an account of the *covariation*.[9] These considerations motivate the following definition:[10]

> *Criterion C:* A *kind concept* is represented as a set of convex regions in several domains together with information about how the regions in different domains covary.

Within this framework, objects are seen as instances of concepts and are mapped into points of the space which determines its properties via the corresponding regions the point belongs to.

Let us next give some formal structure to these ideas. The representation $C(M)$ of a kind concept M can be seen as a subset of the Cartesian product of n domains $D_1, \ldots, D_n$:

$$C(M) \subseteq D_1 \times D_2 \times \ldots \times D_n. \tag{2.2}$$

This subset is nothing more than a collection of points representing possible objects within that concept. This means that any object x falling under M can be represented as an n-dimensional point. Each $x = \langle x_1, x_2, \ldots,$

$x_n\rangle \in C(M)$, represents the coordinates of x in the domain D_i, which will typically fall under some subregion $R_i \subseteq D_i$, corresponding to a subordinate property or concept of D_i.

The central notion of *conceptual space* is defined as a collection of one or more domains with a distance function (a *metric*) that represents properties, concepts, and their similarity relations. The distance function can be of different types; the most common one is the Euclidean, but also Manhattan and polar metrics may be appropriate in different contexts (Shepard 1964; Johannesson 2002; Gärdenfors 2014). For instance, within the context of a Euclidean metric, the distance measure will be dependent on weights w_i that represent the *salience* of dimension i in the conceptual space. Formally the Euclidean distance between two points x and y is defined as

$$d(x,y) = \sqrt{\sum_i w_i (x_i - y_i)^2}\,. \tag{2.3}$$

Here, x_i and y_i are the coordinates of the points in dimension i. We return to a more detailed analysis of salience in section 2.6. Distances are inversely related to *similarity*: The closer two points are in a space, the more similar the entities they represent are. Similarity will be further discussed in section 2.7.1.

For example, consider a basic conceptual space for FRUIT defined as a composition of properties of fruits from the domains of color, taste, ripeness, texture, and shape. Let the "fruit space" be the Cartesian product of these five domains. The concept APPLE will occupy specific subregions of these domains that correspond to the possible properties of these fruits together with correlations between regions of these domains, as it is represented in figure 2.4.

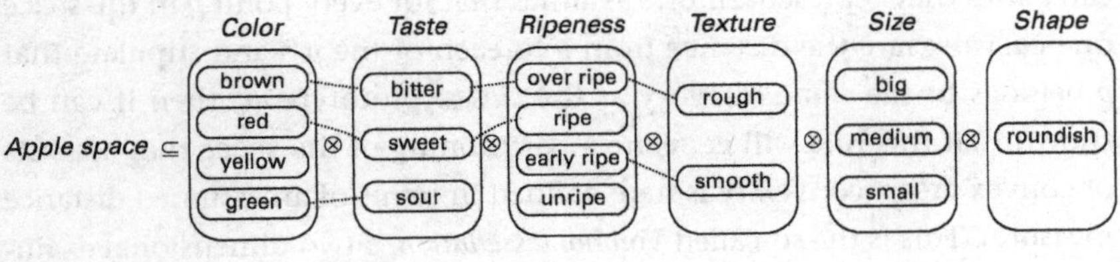

Figure 2.4
The concept of an apple as a subregion of "fruit space." The dotted lines represent covariations between properties for the concept apple.

2.5 Prototypes and Hierarchies of Concepts

2.5.1 Prototypes

Criteria *P* and *C* allow for a natural representation of the prototype effect (Rosch 1975, 1978; Mervis and Rosch 1981; Lakoff 1987; Gärdenfors 2000). Within convex regions, one can take some specific point (or set of points) as the prototype of a category.[11] For example, focal colors are generally considered in cognitive science and linguistics as prototypes of the color space (Rosch 1971; Douven and Gärdenfors 2019). In section 2.7.2, we will use the built-in metric of the space to define a measure of *typicality* of any category member by estimating its distance from the prototype. There are, however, domains without prototypes: In one-dimensional domains such as length, weight, or temperature, one cannot find prototypes of *long, heavy,* or *warm.*[12]

Assuming the prototypical structure of concepts does not require that there is an actual object that represents the prototype—it is just a possible object. Describing concepts as convex regions of a conceptual space allows us to represent the complete space of possible objects that would fall under the concept. In particular, a prototype may be represented as a partial vector, where only the values of the most relevant dimensions for the concept have been determined. For example, the general shape of the prototypical bird would be included in the vector, but its color or age presumably would not.[13]

It is possible to argue in the converse direction, too, and show that, if prototype theory is adopted, then the representation of concepts as convex regions is to be expected. To obtain a prototypically structured conceptual space, we start from a set of prototypes $p_1, \ldots, p_n$ of the categories to be represented (e.g., different bird species). These are the central points in the categories they represent. If one assumes that for every point p in the space one can measure the distance from p to each of the p_i's and stipulate that p belongs to the same category as the *closest* prototype p_i, then it can be shown that this rule will generate a partitioning of the space that consists of convex areas (convexity is here defined in terms of an assumed distance measure). This is the so-called *Voronoi tessellation*, a two-dimensional example of which is illustrated in figure 2.5.[14] Thus, assuming that a metric is defined on the subspace that is subject to categorization, a set of prototypes

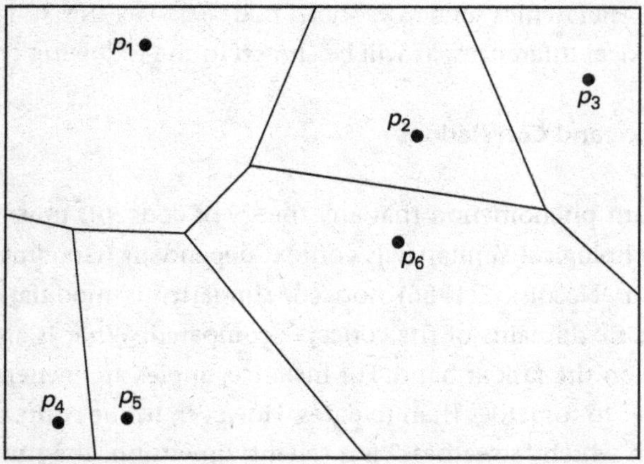

Figure 2.5
Voronoi partitioning in a two-dimensional space.

will, by this method, generate a unique partitioning of the subspace into convex regions.

The method allows for representing graded membership and degrees of typicality (Rosch et al. 1976; Hampton 2007); that is, we can represent objects in the space as being more or less typical instances of the categories according to their position relative to the prototype. Representing typicality in this way has several advantages in terms of cognitive economy for processes like categorization (Gärdenfors 2000, 123–126, 2014, 2024b) and inductive inference, as we will show in chapters 5 and 7 (Gärdenfors and Stephens 2018; Osta-Vélez and Gärdenfors 2020). It is also crucial for representing expectations.

2.5.2 Hierarchies of Concepts

Concepts can be understood as organized in hierarchical structures with a *horizontal* and a *vertical* dimension (Rosch 1988). The horizontal dimension concerns contrast relations between concepts at the same level of abstraction. For example, DOG, CAT, LION, and HORSE are in a contrasting relation since any object falling under one of them is automatically excluded from the others. This type of relation occurs when concepts are included in the same partitioning of a superordinate category (in our example, MAMMAL). The subordinate/superordinate relation is encoded in the vertical dimension. Hierarchies such as <POODLE, DOG, MAMMAL, ANIMAL, LIVING THING> generate lexical inferences, as will be shown in the following chapter.

2.6 Salience and Correlations

An important phenomenon that any theory of concepts must account for is that psychological similarity is context-dependent (Goodman 1972). In particular, as Nosofsky (1986) noticed, similarity is modulated by *attention* to specific domains of the concepts compared. What is attended to is dependent on the task at hand. For instance, apples are (generally) seen as more similar to tomatoes than to dates. However, in the context of making a dessert, in which "sweetness" is a salient dimension, the similarity judgments are expected to change.

Context effects have been extensively studied in the psychological literature (Goldstone et al. 1997; Keßler et al. 2007), and geometrical models of similarity have been often criticized because of their limitations in accounting for such effects (Tversky 1977; see Decock and Douven [2011] for a review).

The conceptual spaces model, however, does not suffer from these shortcomings (Johannesson 2000, 2002). The context-sensitive character of psychological similarity is accounted for in terms of a weighted distance measure, as presented in section 2.4. As noted there, the distance measure includes weights w_i that modify the salience of dimension i in the conceptual space. When a larger value is given to a weight w_i, the conceptual space is magnified along that dimension, which means that dimension i will become more important when determining the similarity between categories (Gärdenfors 2000, 20). As we will show later, this weighted distance function will have a central role in accounting for the role of context in category-based induction.

2.7 Measures of Similarity, Typicality, and Diagnosticity

In this section, we show how the distance function of a conceptual space, together with some notions from prototype theory, can be used to define formal measures of similarity, typicality, and diagnosticity. In chapter 9, we will present a measure of the coherence of a concept. These notions are central to psychological theories of reasoning with concepts, but in the literature they have not been defined formally. We believe that the measures we propose are one of the most novel contributions of our book. Since they are all based on the theory of conceptual spaces, they show the fruitfulness of our framework. The measures will be central to our models of the various forms of reasoning.

2.7.1 Similarity

As previously stated, in categorization research, it is often assumed that the *similarity* of two stimuli can be determined from the distances between the representations of the stimuli in the underlying psychological space. The smaller the distance between the representations of two objects, the more similar they are. Consequently, conceptual spaces provide us with a natural way of representing similarities. What is then this functional relation between similarity and distance? In section 2.3, we presented Shepard's formula for calculating similarities from distances. This equation can be inverted to a formula for determining distances from similarity. As before, let $sim(i, j)$ express the similarity between two objects i and j, where $0 < sim(i, j) \leq 1$, and $d(i, j)$ their distance, then the following formula is the inverse of equation 2.1:

$$d(i,j) = -\frac{1}{c}\ln sim(i,j). \qquad (2.4)$$

Tversky (1977) has criticized metric accounts of similarity. In particular, he focuses on the fact that metrics are symmetric—that is, the distance from x to y is the same as the distance from y to x, which also makes the similarity relation symmetrical. In a number of experiments, he and his colleagues have demonstrated that similarity judgments are often asymmetrical. For example, Tel Aviv is judged to be more similar to New York than New York is similar to Tel Aviv. As an alternative to spatial models of similarity, Tversky (1977) proposed a set-theoretic model based on property matching or feature matching, which can account for the violations of the distance axioms.

Because Tversky assumes that the similarity of objects is determined from fixed sets of features for the objects, it seems that his criticism presumes that similarity is *not* dependent on context. A better way to explain the experimental results is by considering the cognitive *process* underlying the similarity judgments (Medin et al. 1993, 258). In criterion C, the salience of different dimensions plays an important role. This means that similarity judgments are dependent on the weighting of dimensions. So, for example, when Tel Aviv is compared to New York, other dimensions will become prominent than when New York is compared to Tel Aviv. As Nosofsky (1992, 40) pointed out, a plausible cognitive process model of the subject's similarity judgment is that subjects assign greater salience to those dimensions along which stimuli are more similar. Hence, subjects actively search for ways stimuli are similar. Thus, criterion *C* can also handle the alleged violations of the triangle inequality by using different patterns of salience weights for different comparisons.[15]

2.7.2 Typicality

Because conceptual spaces are metric structures, they allow us to measure the *typicality* of properties, which is a central variable for reasoning with concepts. The core idea behind our proposal is that the typicality of a property *R* for concept *M* can be estimated by measuring the distance from the prototype of *M* to the nearest counterexample of *R*.[16] For example, in most contexts for APPLE, *juicy* is more typical than *red* because the nearest nonred apple lies closer to the prototype of APPLE than the nearest nonjuicy apple.

To develop this idea more formally, let a concept *M* be represented by a region in a conceptual space $C(M)$ with prototype $p_M \in C(M)$. For any point $x \in C(M)$ and any property $R \subseteq D \subseteq C(M)$, "$R(x)$" means that the coordinate of *x* on dimension *D* falls within the subregion *R*. For the definition of typicality, we must distinguish *prototypical* properties *for M*—those for which $R(p_M)$ *holds*—from *nonprototypical* properties—those for which it does not ($\neg R(p_M)$). While prototypical properties are evaluated by the nearest counterexample, the typicality of a nonprototypical property is determined by the distance from the prototype to the nearest example.

We therefore define the typicality of a property *R* in $C(M)$—written $T_M(R)$—as follows:

Typicality measure:

$$T_M(R) = \begin{cases} \min\limits_{x \in C(M)-R} d(x, p_M), & \text{if } R(p_M) \ (R \text{ is prototypical}), \\ -\min\limits_{x \in R} d(x, p_M), & \text{if } \neg R(p_M) \ (R \text{ is nonprototypical}). \end{cases} \tag{2.5}$$

From this definition it follows that, for any two properties R_i, R_j in $C(M)$, $T_M(R_i) > T_M(R_j)$ if and only if $T_M(\neg R_j) > T_M(\neg R_i)$. Furthermore, since nonprototypical properties are always assigned negative values, they will be less typical than all the prototypical properties. Values close to zero indicate borderline cases: the prototype lies almost as close to the nearest instance x with $R(x)$ as to the nearest counterexample y with $\neg R(y)$.

It is important to note that in our typicality measure, we do not count the number of instances, but the criterion is based on similarity to the prototype. In other words, our model is not probabilistic.[17] Probabilistic models will not give the right results for typicality since some properties that are probable may be very atypical.

This measure is closely related to the notion of "mutability" of an attribute (Kahneman and Miller 1986, 137; Love and Sloman 1995; Sloman et al. 1998). Mutability is the ease with which a feature (property) of a concept can be changed while maintaining its identity. For example, *has a red breast* is a mutable feature of ROBIN, as we can easily imagine a robin without it and still consider it a robin. In contrast, *has wings* is an immutable feature, as it's much harder to imagine a wingless bird and still call it a robin. As Sloman et al. (1998) show, mutability is closely related to coherence. Our measure of typicality, however, can be defined independently of this notion. We will explain the relationship between them in chapter 9.

2.7.3 Diagnosticity

As explained in the previous chapter, a property is diagnostic when it maximizes the dissimilarity between a category and the categories in its contrast class, speeding up categorization (Tversky 1977).[18] In our model, we address the influence of the contrast class by considering the prototype of the immediate *superordinate* concept. Consider the concept SUBMARINE and its superordinate category VEHICLE. The property *operates underwater* is highly diagnostic for submarines because it maximizes their dissimilarity from any other instance of vehicle; by contrast, *transports people or cargo* is common across VEHICLE and thus not diagnostic for submarines.

We can define diagnosticity in terms of typicality in conceptual spaces as follows:

(1) A property R is *diagnostic* for concept M—relative to its immediate superordinate N—if R is not a prototypical property of N.

(2) Given concept *M,* its superordinate *N,* and property *R* in *C*(*N*), we define the diagnosticity of *R*,—written $D_{M,N}(R)$—as follows:

$$D_{M,N}(R) = \frac{T_M(R)}{T_N(R)}. \tag{2.6}$$

Diagnosticity is related to typicality but not the same: Diagnostic properties are typical, but not all typical properties are diagnostic. For example, *sweet* is typical of APPLE but not diagnostic, since several categories in the contrast class share it.

2.8 Neuroscientific Correlates

The first thing to note about the brains of humans and other mammals is that the cortex abounds in topographic maps, whereby neighborhood relations at the sensory periphery are preserved in the arrangement of neurons in various regions of the central nervous system. For example, one finds retinotopic maps in the lateral geniculate nuclei, which are arranged in six layers, each layer arranged in a topographic representation of the retina; there are somatotopic maps representing sensory positions on the body; and there are tonotopic maps, which transform the pitch dimension of a sound to a spatial dimension along the cochlea. In these maps, the orderly mapping of neurons to sound frequencies is preserved from the cochlea to the auditory cortex. The topographic maps correspond to distinct domains of representation since most of them preserve the modularity of the senses. The underlying reason is that distinct types of receptor neurons are sensitive to different features of our environment, and these features are kept distinct in the maps higher up in the projection system of the brain.

More recently, the hippocampal system—that is, the hippocampus together with the entorhinal cortex—has been proposed as a system for representing not only locations in ordinary space but for representing different kinds of conceptual spaces (Constantinescu et al. 2016; Bellmund et al. 2018). According to this view, such spaces rely on the activity of spatially tuned cells in the hippocampal system. The dynamic mapping system enables rapid reorganization of "coordinate systems" through remappings between orthogonal representations across different conceptual domains. The result is a multitude of cognitive spaces at different resolutions and hierarchical levels.

During navigation in ordinary space, place cells in the hippocampus show increased firing at a particular location within the space. Then, grid cells in the adjacent entorhinal cortex fire at multiple locations within an environment, and these locations form a hexagonal grid. Combining the information from place cells and grid cells, these cells generate self-localization. The coding also supports geometric computations when planning spatial navigation by representing distances and directions.

Several studies have shown that similar mechanisms are exploited in tasks involving other conceptual spaces (e.g., Caglar 2021; Zheng et al. 2024). In an experiment by Aronov et al. (2017), rats manipulated the frequency of an auditory stimulus to "navigate" to a target frequency. Cells in the hippocampus and entorhinal cortex showed firing fields constrained to each cell's preferred frequency range, thereby mapping the dimension of sound frequency so that neighboring frequencies exhibited similar firing patterns across cells.

Another example is presented by Nitsch et al. (2024), who study how a two-dimensional value space used in a decision task becomes represented in the brain. The result is that the entorhinal cortex exhibits a grid-like representation with an orientation aligned to the axis through the value space that is most relevant for choices. These experiments suggest that the grid cells function as a "universal coordinate system" that can be used to represent a number of different conceptual spaces.

2.9 Implementations of Conceptual Spaces

Conceptual spaces have proven to be a versatile and powerful tool for addressing a wide range of problems in multiple fields. For example, they have been used as a tool for representing and organizing information in disciplines as varied as geography, computer science, and political science (Ahlqvist 2005; Song and Bruza 2001; Laver 2014).[19]

In this book, we will not directly consider computational applications of the model. However, the possibility of developing such applications is one of our motivations for presenting models based on distance measures in conceptual spaces. Since there already exist computational models of conceptual spaces (Chella et al. 2001; Adams and Raubal 2009; Schockaert and Prade 2013; Lieto 2021), our accounts of the different forms of concept-based reasoning can, in principle, be computationally implemented. For example, in chapter 8, we will present semi-algorithmic methods for handling four different types of composed analogies.

Within AI, researchers have noticed that the problem of the format of representation of knowledge is crucial for modeling processes like reasoning in artificial systems (Lakemeyer and Nebel 1994; Lieto 2021). The propositional view, widely popular at the beginnings of AI thanks to the prevalence of logical models, lost popularity for a simple reason: Classical logic has several computational disadvantages when used as a framework for knowledge representation (Minsky 1991; Sowa 1991; Lifschitz, et al. 2008).

The main competing model for reasoning with concepts is Bayesianism (Oaksford and Chater 2007; Tenenbaum et al. 2006, Tenenbaum et al. 2007). From a computational point of view, however, computations with such models would be very complicated. The reason is that Bayesian models presume very rich information concerning probability distributions over properties and combinations of properties (Cooper 1990; Kwisthout et al. 2011; Bowers and Davis 2012). We find that the relative ease of implementation is a strong argument for models based on conceptual spaces. This said, we do not claim that computer models are easy to develop.

The first step for such implementations would be to learn the geometric structure of a domain. Wheeler and Natarajan (2024; see also Bouraoui et al. 2022) present a computational method for extracting domains from a dataset. They build on three requirements for the domain learning process:

(1) A model mapping data to interpretable dimensions.
(2) A set of prototypes defining property regions.
(3) Semantically consistent similarities between properties.

To satisfy these requirements, they propose an autoencoder-based approach. It would take us too far to present their method here, but their experiments with the MNIST dataset of images of handwritten digits and the CelebA dataset of color images of celebrity faces show promising results.

The second step would be to describe *domain structures*. This involves, above all, specifying their geometric and topological structure. The third step is to give information about how the resulting space is partitioned into concepts. Using prototypes and Voronoi tessellations, this can be done in a computationally efficient way, once the relevant metric is determined.

Representing information by conceptual spaces requires computations that involve vectors, using inferences based on similarities rather than mechanisms based on tree searching as is standard in a rule-based symbolic approach. Developing such implementations of conceptual spaces could enable the creation of new forms of automated reasoning that go beyond systems based on logical formalisms or neural networks, something that would be useful in the efforts to simulate commonsense reasoning with concepts.

In the theory of conceptual spaces, the notion of convex regions plays a central role (Gärdenfors 2024). Tětková et al. (2024) investigate to what extent convexity is satisfied in machine-learned latent spaces—for example, artificial neural networks generated by deep learning. First, they show that convexity is robust to reparameterizations of a network. This basically depends on the fact that convexity is conserved under affine transformations. Then, they studied the classification generated by databases containing images, audio, human activity, and brain data. The results were that convexity is approximately satisfied in the representations that are generated by the artificial neural networks. They conclude that the degree of convexity "is an indicator of how well pretrained networks are prepared for future generalizations by the self-supervised learning procedure."[20]

A related study was performed by Steinert-Threlkeld and Szymanik (2020). They generated three hundred different artificial color systems by partitioning the CIELab color space. For each color system, they measured their degree of convexity. They then trained an artificial neural net system to learn to color categories of each system and measured how *learnable* each system is. Their result was that there is a fairly strong correlation between the degree of convexity of a color system and its learnability (see their figure 5).

Similar results have also been obtained by Douven and Verheyen (2024). Like Steinert-Threlkeld and Szymanik (2020), they generated artificial color and shape categorizations. The system either consisted of convex or connected regions. Also, using a computational system to simulate the learning process, they demonstrated that convexity facilitates learning more than does connectedness. Taken together, these two studies give independent support for the claim that convex concepts are indeed easier to learn than concepts that are not convex.

This line of research is continued by Dorszewski et al. (2024), who investigate the connection between concept convexity and how well human judgments align with classifications in deep neural networks. Their experiment is based on the THINGS dataset, which contains 26,000 images of 1,854 classes of natural objects grouped into twenty-seven high-level concepts. For these concepts, human judgments of similarity have been obtained to analyze the degree of alignment between the human and the network representations of concepts. Dorszewski et al. (2024) show that, generally, one finds a high correlation between the degree of alignment and the convexity of the twenty-seven high-level concepts. This suggests that the neural networks form convex regions in their latent spaces.

Interestingly, large language models (LLMs), such as Chat-GPT and Bard, can potentially be used to generate domains and their structure. If one, for example, asks Chat-GPT to judge the similarity between two mammal species, the replies are often sensible and can be used as input to methods of multidimensional scaling or principal component analysis. This would be a much cheaper and more efficient method to generate domains of conceptual space instead of letting human subjects make similarity judgments (Moullec and Douven 2025). The drawback is, of course, that the data generated by large language models are less reliable than human data, but the method could function as a first approximation.

Another investigation of the underlying conceptual structure of LLMs is presented by Patel and Pavlick (2022). They show that LLMs can learn to map an entire conceptual map even without being grounded in the external world. One example they study is the color space, and they show that GPT-3 can generalize from training data from the RGB color space to "unseen" colors. However, even if GPT-3 can learn the relations between the different shades of the color domain, the set of relations is still "ungrounded" in the sense that if the space was rotated 180° along the color circle, the mapping would still be the same. There is, therefore, no connection between the "green" area of the RGB representations and green things in the world.

In an investigation of ChatGPT and GPT-4, Chatterjee and Schockaert (2023) found that the LLMs have access to rich conceptual information, which makes it possible to identify similarity structures (see also Chatterjee et al. 2023; Kumar et al. 2024; and Liu et al. 2025). They suggest that "recent LLMs have achieved a level of conceptual understanding that goes qualitatively beyond what was possible with earlier models" (2023: 3).

Also, Moullec and Douven (2025) investigate to what extent LLMs can be used to create "cheap" conceptual spaces—that is, spaces that can be generated without costly and time-consuming psychological tests collecting similarity judgements from human participants. They use the set of mammal species that was used in Henley's (1969) three-dimensional estimation of "mammal space" on the LLMs BERT and Llama 3.2.[21] The results are disappointing in the sense that the LLMs gave results that fit worse to human judgement than those that were achieved by standard natural language processing systems such as Word2Vec and FastText. These results indicate that the possibility of using LLMs to generate spaces for computational applications should be taken with several grains of salt. The development in the area is fast, though, so better methods may be around the corner.

Computational implementations of conceptual spaces are still at an early phase. We believe, however, that the models presented in this book will be useful for developing artificial systems capable of reasoning with concepts in the areas we address. The methodology for doing this will, however, be different from the techniques currently used to create LLMs.

2.10 Conclusion

In this chapter, we have introduced the theory of conceptual spaces that will form the framework for our models in the following chapters. Further details about the theory can be found in Gärdenfors (2000, 2014). With the aid of the metric associated with a conceptual space, we have defined measures of similarity, typicality, and diagnosticity that will form the cores of the models we develop in the following chapters.

3 Lexical Entailment in Conceptual Spaces

In this chapter, we examine a type of inference that seems deductive, since it involves no uncertainty, but does not fully conform logical rules. We call it "lexical" since it relies on identifying *entailment relations* among lexical concepts (see Chierchia and McConnell-Ginet 1990, 361–362; Cummings 2005, 80–84). Our goal is to show, through some examples, how these inferences can be explicated by the formal properties of the conceptual spaces underlying the premises and the conclusion. In the chapters that follow this one, we will use the same strategy to analyze forms of inference that involve uncertainty.[1]

In the remainder of this chapter, we discuss how lexical inference has been treated within the formalist tradition, focusing on the notion of *meaning postulates*. Then we turn to decompositional models, which explain lexical entailment through *semantic primitives*. Finally, we introduce our own model based on conceptual spaces, showing how it provides a more systematic and psychologically grounded explanation of lexical inference.

3.1 Two Models of Lexical Entailment

3.1.1 Meaning Postulates

According to the formalist tradition, the rules that guide reasoning are sensitive only to the syntactic properties of propositional attitudes (i.e., beliefs, assumptions, and so on) that figure in arguments. Since these syntactic properties are thought to mirror semantic ones (such as the truth-conditions of propositions), they allow reasoning to preserve truth and, when the attitudes are accurate, to represent states of the world correctly (Fodor 1985, 22–23). In other words, syntactic derivations within the system correspond to semantic entailments. In this framework, the inference of (3.1b) from (3.1a) and the inference of (3.2b) from (3.2a) are both instances of the same reasoning pattern and thus receive the same psychological explanation: They are accounted for by the rule of conjunction elimination.

(3.1) (a) John opened the door and entered the room.

(b) John entered the room.

(3.2) (a) The table is big and round.

(b) The table is big.

Any competent English speaker would recognize (3.1) and (3.2) as valid inferences purely from their grammatical structure (see Cann 1993, chap. 3). Yet some equally intuitive inferences resist such logical analysis, posing a challenge to the formalist strategy. Consider now the following pairs of sentences:

(3.3) (a) John killed Jason.

(b) Jason is dead.

(3.4) (a) Mary is John's mother.

(b) Mary is older than John.

What logical rule explains why we see (3.3) and (3.4) as evidently valid inferences? The answer is none—at least no rule of standard logic. These inferences depend not on syntactic properties but on our knowledge of the nonlogical vocabulary of English, namely, our *lexical knowledge*. In other words, in this class of inferences, which are prevalent in everyday reasoning, the semantic properties in question have no parallel in the syntactic structure of propositions.[2]

It is an important challenge for logicians and linguists studying natural language inference to develop a notion of entailment that captures both formal and lexical inferences (Cruse 1986, 13–14; Cummings 2005, chap. 3; Partee 2011). Montague (1973), building on Carnap (1952), offered the first solution to this problem. He proposed to "enrich" the formal logical system for modeling natural language entailment by adding nonlogical axioms, called "meaning postulates," which link terms in the nonlogical vocabulary thereby restricting the set of models that can render a statement true. For instance, the meaning postulate "If x is a man, then x is an animal" limits the admissible models for the sentence "John is a man" to those that also admit the sentence "John is an animal."

Meaning postulates offer an *extensional* solution—one that is compatible with a deductivist and formalist views of reasoning—to the problem of characterizing lexical meaning and lexical entailment (Chierchia and McConnell-Ginet 1990, chap. 2.2).[3] We can account for our intuitions about the validity of inferences (3.3) and (3.4) by assuming that our knowledge of the lexical concepts "kill," "die," "mother," and "older" also includes knowledge of the meaning postulates $\forall x \forall y(Kill\ (x, y) \rightarrow Dead(y))$ and $\forall x \forall y\ (Mother\ (x, y) \rightarrow Older(x, y))$. Presumably, these postulates are accessed and applied during the semantic interpretation of the premise, which licenses the conclusion and renders the inference logically valid.

Originally developed in formal semantics (Zimmermann 1999), meaning postulates were later imported into psychology. Fodor (1975) suggested that lexical meanings in natural language could be modeled as meaning postulates within a mental language that supports nonlogical inference. However, this approach to lexical meaning and entailment has significant limitations.

First, as Marconi (1997) pointed out, our lexical competence extends far beyond our inferential activity. Lexical concepts allow us to categorize objects and situations in the world, make distinctions, answer questions, and guide action in various ways. In short, the informational content of lexical meanings is not exhausted by its semantic interpretation-cum-meaning postulates. Second, this approach suffers from what Harnad (1990) called the "symbol grounding problem." Roughly, it fails to explain how nonlogical terms of natural language relate to extralinguistic elements since they are characterized in purely intralinguistic terms. As Searle's (1980) Chinese room argument has shown, it is difficult to say that someone genuinely "understands" a language by merely manipulating uninterpreted symbols according to rules. In a similar vein, Partee (1981, 71) writes that "no amount of such intra-linguistic connections [meaning postulates] can serve to tie down the intensions with respect to their extralinguistic content. For that, there must be some language-to-world grounding."

Furthermore, several semantic phenomena related to the content of lexical concepts cannot be explained in terms of nonlogical axioms like meaning postulates. For example, the entailment relation between the lexical concepts "bird" and "fly" is not the same as that between "bird" and "animal." The former licenses "risky" nonmonotonic inferences (e.g., "penguins are birds but do not fly"), while the latter allows inferences without uncertainty. Meaning postulates do not capture the prototypical structure of concepts and how this affects their entailment relations (see chapter 5).

Finally, it is not possible to capture higher-order semantic relationships between sentences through meaning postulates. For example, we all understand the analogy "The hand is to the arm what the foot is to the leg" (see chapter 8). However, the meaning of this statement cannot be explained in terms of truth-conditions. Moreover, the analogical relationship between its lexical components cannot be explained in terms of (first-order) meaning postulates.

3.1.2 Decompositional Models

The rationale for explaining lexical meaning and entailment in terms of meaning postulates is that it allows us to stay at the sentential level of semantic analysis, committed only to truth conditions and extensional structures. However, the problems of this strategy indicates the need for a deeper, subsentential level of semantic analysis that incorporates a *lexical* perspective. Decompositional approaches in semantics aim to address this need (Chierchia and McConnell-Ginet 1990, chap. 2; Jackendoff 2012).

The idea behind the decompositional view is that lexical meanings are not unanalyzable atomic units but can be broken down into basic "components of meaning" or "semantic primitives" such as CAUSE, BECOME, ACT, POSS (possession), and LOC (location), along with argument roles like *agent*, *patient*, *recipient*, and *instrument* (Fillmore 1982; Jackendoff 1990; Levin and Rappaport Hovav 1994; Gärdenfors 2024a). By decomposing word meanings in this way, the theory can capture systematic relationships between semantically related words and account for regularities in their syntactic behavior. For instance, the semantic similarity between the words "kill," "die," and "dead" can be explained by positing that they all contain the primitive DEAD in their decomposed representations, with "kill" additionally including a CAUSE predicate (Wunderlich 1997).

Decompositional models aim to provide systematic explanations of patterns of lexical entailments. For example, in (3.3a-b), the verb "kill" can be decomposed as [*x* CAUSE [BECOME [*y*<DEAD>]]], meaning that there is a causing event performed by *x* that resulted in *y* becoming dead (Wunderlich 1997). The verb "die" can be decomposed as [BECOME [*y*<DEAD>]], expressing the change of state to being dead without specifying a cause.

Given these representations, "kill" clearly entails "die" since the decomposition of "kill" properly includes the decomposition of "die."

Similarly, the semantics structure of ditransitive verbs like "give" can be explained by the decomposition [*x* CAUSE [BECOME [*y* HAVE *z*]]], where *x* causes *y* to come to possess *z* (Pinker 1989; Krifka 1999). This representation involves a change of possession, so "give" entails "receive." It also entails that *x* causes this change of possession, so "*x* gives y to *z*" entails that "*x* causes *z* to receive *y*."

Jackendoff's conceptual semantics explains how different patterns of lexical entailment can arise from the same decompositional structure. For example, the three cases of entailment with causative verbs shown in table 3.1 instantiate the single general pattern: X_{cause}[E to_occur] → E occurs.

This explanatory strategy—which can be generalized to other lexical types—helps explain why we see different lexical entailments as structurally similar. Furthermore, it can explain various "sense relations" relevant to natural language inference, including hyponymy, antonymy, anomaly, ambiguity, polysemy, homonymy, and contradiction (Cruse 2000). More broadly, we think that decompositional theories are better equipped than purely truth-conditional approaches to account for our lexical-inferential competence, while also offering a more "natural" bridge to psychological theories of linguistic competence.

Conceptual spaces can be broadly understood as a decompositional approach, but one that departs from other frameworks in cognitive semantics.The main difference is that instead of postulating fixed semantic primitives, it builds on the more flexible notions of dimension and domain (see section 2.2), which allows the theory to connect directly with ideas and results from experimental and theoretical psychology. Another important

Table 3.1
Inferential patterns for causative verbs

Sentence	Decompositional structure
X killed Y → Y died	*X kill Y*: X_{cause}[*Y die*]
X lifted Y → Y rose	*X lift Y*: X_{cause}[*Y rose*]
X gave Z to Y → Y received Z	*X gave Z to Y*: X_{cause}[*Y to receive Z*]

From Jackendoff (1992, 39).

strength is its precise and testable account of *semantic similarity*, which not only helps explain how lexical information supports inference but also fits well with experimental methods that investigate similarity spaces.

3.2 Lexical Inference in Conceptual Spaces

3.2.1 Core meaning, Attention, and Conceptualization

An advantage of conceptual spaces over other theories in cognitive semantics is that they can provide *algorithmic explanations* (Marr 1982; Love 2015) of semantic phenomena. An algorithmic explanation characterizes a cognitive process in terms of the representations on which it is based and the operations that transform these representations to generate an output from a given input. This kind of explanation bridges the gap between abstract computational descriptions and concrete neural implementations. In the case of conceptual spaces, the theory posits a specific representational framework and formally describes operations on those representations that explain different aspects of our semantic competence. In other words, this approach can provide a mechanistic understanding of semantic competence, linking intuitive semantic phenomena to concrete, computationally specifiable operations. Later in the chapter and in the following chapters, we will illustrate this through multiple examples.

The first problem we face when analyzing lexical inference is establishing a criterion for distinguishing the kind of conceptual content that supports inferences that are genuinely "lexical" from other types inferences. For instance, possessing the concept DOG requires knowing the concept MAMMAL and properties such as *four-legged* and *domestic* (among many others) but not necessarily concepts like SESAMOID BONE or VOMERONASAL ORGAN.

Kiefer (1988) distinguishes three layers of lexico-conceptual knowledge. The first is *linguistic knowledge*, which constitutes the *core meaning* of a lexical item. The second is general *conceptual knowledge*, which concerns "predictable modifications of the core meaning in various contexts" (ibid., 1). The third is *encyclopedic knowledge, which* goes beyond linguistic competence and is related to broader cultural or experiential expertise. Determining the boundaries between these three layers is a philosophically contested matter (see Marconi 1997, chap. 2; Paradis 2003), one with direct implications for how we specify the scope of lexical inference.

To avoid the above problem, we restrict lexical inference to inferences based on knowledge of core meaning. These inferences involve no uncertainty from the agent's perspective: Any competent language user treats them as totally warranted by meaning. This entails that, if an agent makes a lexical inference and later learns it was incorrect, she will likely revise her knowledge of the concept that supported it. Consider, for example, a child who, upon seeing a dolphin, infers that she is seeing a fish. If someone corrects her and explains her that dolphins are mammals, she will revise her concept, recognizing that not all swimming creatures belong to the same category. This adjustment involves reclassifying DOLPHIN within her animal taxonomy—an instance of conceptual change. This "revision criterion" is important in our analysis; in later chapters, we will extend the discussion to concept-based inferences that can be revised without inducing changes in the agent's conceptual system.

Our analysis of lexical inferences draws some inspiration from cognitive semantics, where a central tenet is that meaning is a form of *conceptualization*: Lexical items and sentences evoke mental representations in the minds of speakers/hearers, rather than merely serving as mappings onto a language-independent, external, reality (Fillmore and Atkins 2000; Langacker 2000). Consider the lexical concept "Tuesday." Its semantic profile is not determined by some relation to a mind-independent entity but by a conceptual construct, organized through our mental and cultural understanding of time and its relationship to other concepts like "day," "week," and "month." With respect to sentential meaning, we find a conceptualization pattern in which sentences feature a *designatum* that "stands out" against a *background* or *base* of associated concepts. These structures can be organized in various ways, leading to different "construals" that allow a single situation to be conceptualized differently. Langacker (2002, 61) illustrates this idea in the following way:

> A speaker who accurately observes the spatial distribution of certain stars can describe them in many distinct fashions: as a *constellation*, as a *cluster of stars*, as *specks of light in the sky*, etc. Such expressions are semantically distinct; they reflect the speaker's alternate construals of the scene, each compatible with its objectively given properties.

This linguistic phenomenon is attention-driven (Lampert 2009; Talmy 2007). Cognitive linguists often draw parallels with visual perception to show how attention shapes conceptualization (Langacker 1987, 116;

Gärdenfors 2014, section 1.3.3). Their point is that, just as we rely on attention-driven figure-ground segregation to navigate our physical environment, we use similar mechanisms when processing language.[4]

The construal of a sentence imposes a *profile* onto a conceptual base, highlighting certain elements while disregarding others. For example, the sentence "John is painting the door green" profiles an action performed by an agent. The same event can be *reprofiled* as "The door is being painted green by John," shifting the designatum while preserving the underlying conceptual base (Langacker 2009, chap. 1). This process of *reprofiling*—or "alternate profiling" in Langacker's terms—involves an attentional shift within a conceptual base, producing minimal semantic transformations. We claim that lexical inferences can be understood as cases of reprofiling sentences within their associated conceptual structures. For example, the inference "If something is a dog, then it is a mammal" reprofiles an object from the conceptual space of DOG toward that of MAMMAL, which is implicit in the representation of DOG as a subregion of MAMMAL.

3.2.2 Kind Concepts

We start by discussing lexical inferences with kind concepts. According to the conceptual space model, kind concepts are represented by convex regions in a conceptual space—that is, a subset of the product space of the set of dimensions that constitute the space. For a kind concept M, $C(M)$ corresponds to a subset of the Cartesian product of n domains. As explained in the previous chapter, an object x falling under M corresponds to an n-dimensional point $x = \langle x_1, x_2, \ldots, x_n \rangle \in C(M)$ with the coordinates of the point in each dimension of the space.

In most cases, x is a partial vector[5] since specific values on some dimensions are often unknown. This lack of information creates uncertainty when reasoning with these points. The following chapters argue that everyday reasoning addresses these information gaps by "filling in" the unknown elements of the vector with values from the prototype of the concept. The resulting inferences are based on *expectations* (see chapter 5 and Osta-Vélez and Gärdenfors [2021]). For example, if told that x is an apple, one might make inferences with varying degrees of uncertainty about its prototypical properties, such as being *round*, *red*, and *sweet*.

Expectation-based inferences derive from our semantic knowledge and can thus be characterized as "lexical." However, they differ from "core" lexical inferences in an important respect: While lexical inferences depend on stable definitional or "essential" properties encoded within lexical concepts (such as "If x is an apple y, then x is a fruit"), expectation-based inferences depend instead on statistical or typical knowledge about categories or situations. As a result, they operate under uncertainty, tolerating exceptions and easily yielding to counterevidence. In contrast, as we said earlier, challenging a core lexical inference requires revising the underlying conceptual structure, as it is not a matter of updating expectations but of restructuring knowledge itself.

Returning to nouns, we distinguish two types of entailment relations that ground lexical inferences. First, there are *upward entailment relations*, where applying a noun M to an object x entails that any other noun N superordinate to M can be equally applied to x. For example, "If x is a cat then x is a mammal." Linguists refer to this kind of relation—*X is a kind/type of Y*—as *taxonomic* (Cruse 2000, chap. 9). From the perspective of conceptual spaces, these relations are grounded in a simple set-theoretical fact: When an entity x is categorized under concept M, it is represented as a point in $C(N)$. Thus, for any concept N such that $C(M) \subset C(N)$, if $x \in C(M)$, then $x \in C(N)$. Due to the transitive nature of set inclusion, this extends to every concept higher than M in the hierarchy.[6] Thus, categorizing x as M will make x (by default) a member of every superordinate concept of M.

Second, there are conceptual relations of *downward entailment*, in which categorizing an object under the noun N licenses the application of other predicates from the internal structure of the concept represented by N. Examples of this are social categories like "bachelor," that licenses lexical inferences such as $Bachelor(x) \rightarrow (Unmarried(x) \wedge Male(x))$. Because the conceptual space associated with the noun "bachelor" extends only into the *unmarried* and *male* regions of the *marital status* and *gender* domains, any person categorized as a bachelor will have coordinates within these regions. Once again, we can see how lexical inference can be explained in terms of the formal structure of conceptual representation.

On the other hand, downward entailment relations with nouns that refer to natural categories (such as BIRD or ROSE) are not that simple because they generally involve uncertainty due to the variation of properties that these categories tend to have in most of their dimensions. We will return to this topic in chapter 5.

Co-hyponyms are lexical terms at the same conceptual level that share one or more common superordinate categories—the *hypernym*. For example, CAT, DOG, RAT, and COW are co-hyponyms with MAMMAL as *hypernym*, and BLUE, GREEN, and RED are co-hyponyms under the *hypernym* COLOR. Co-hyponymy is often considered an *incompatibility relation* (Cruse 2000, chap. 9), meaning that if a co-hyponym applies to an entity, the others at the same level cannot. This gives rise to a type of lexical inference with negation: For instance, the sentence *Dog*(*a*) lexically entails ¬*Cat*(*a*), ¬*S eal*(*a*), ¬*Whale*(*a*), and so on, for all other animal categories that are co-hyponyms of DOG.

Explicating this type of inference in conceptual spaces is straightforward. The set of concepts $N_1, \ldots, N_n$, co-hyponyms of *M*, will be represented as disjoint subregions of the *C*(*M*). This means that $N_i \cap N_k = \varnothing$ for all $i \neq k$. Since the sentence $N_i(x)$ is represented by an object $x \in N_i \subseteq C(M)$, then $x \notin N_k \subseteq C(M)$ since N_i and N_k are disjoint subregions of *C*(*M*). This simple set-theoretical fact justifies all the lexical inferences of the form $N_i(x) \rightarrow \neg N_k(x)$ for every category N_k which is a co-hyponym of N_i.

3.2.3 Relational Concepts and Spatial Prepositions

In this subsection, we briefly discuss lexical inferences based on entailments between relational concepts—that is, concepts about relations among two or more arguments (entities, events, or properties). For reasons of space, we focus only on spatial prepositions, which are terms used for specifying locations, directions, motions, and other space-related states of affairs (Landau and Jackendoff 1993). These prepositions, which are grounded in primitive intuitions about how we represent space and spatial relations among objects (Talmy 1983; Levinson 2003), make an interesting case study because the conceptual spaces underlying them are different from those of nouns and adjectives, which constitute most of the case studies in this book.

Despite being often overlooked in the philosophical literature on concepts and inference, prepositions figure in some classic examples of lexical inferences. As a starting point, we introduce some basic notions from cognitive linguistics used in their analysis. Consider the following sentences:

(3.5) (a) Buenos Aires is *west of* Montevideo.

(b) John went *to* the supermarket.

(c) The cat is *on* the mat.

The three sentences describe an entity in some spatial relation to another entity. The described entity is called a "trajector." It can be static, like "Buenos Aires" in (3.5a) or "the cat" in (3.5c), or dynamic, like "John" in (3.5b). The other entity is called "landmark," and it is the object according to which the trajector's location—or trajectory of the motion—is specified (Langacker 1987). It is important to point out that spatial terms do not describe the exact locations of entities in space. Instead, they give information about a region in which an object might be within a particular spatial configuration. For instance, the sentence "The pen is *on* the desk" would be true if the pen is in the center of the desk, on one corner, or in any other possible position within the table's surface. We will come back to this point later.

Now, the guiding idea behind our analysis of lexical inferences with spatial prepositions is analogous to the one presented in the previous subsection. It builds on the assumption that we conceptualize sentences through representational structures that are exploited in inferences in different ways. Again, we assume that these representational structures are conceptual spaces, and we will directly build on Gärdenfors (2014) for our analysis. Simple examples of lexical inferences can be derived from prepositions that depend on a single dimension. For example, from "The lamp is over the table" it follows that "The table is under the lamp," and from "The castle is east of the church" it follows that "The church is west of the castle."

Unlike what is typical for nouns and adjectives, many spatial prepositions seem to be more naturally represented in conceptual spaces by using a *spherical* or a *polar* coordinate system (Gärdenfors 2014). A spherical coordinate system is a three-dimensional space in which every point p is described via the three parameters $\langle r, \theta, \phi \rangle$, where r is the distance between p and the origin of the system (with $r \geq 0$), θ is the *polar* angle measured from the positive z-axis (with $0° \leq \theta \leq 180°$), and φ is the *azimuth* angle measured in the xy-plane from the positive x-axis to p's projection (with $0° \leq \varphi \leq 360°$) (see figure 3.1).

Using this coordinate system, Gärdenfors (2014) defines the notion of *polar betweenness*, which can be used to define *polar convexity* for concepts in these particular spaces. It should be noted that polar convexity results in different types of regions than when Cartesian coordinates are used (see Zwarts and Gärdenfors [2016] for examples). This notion will show that,

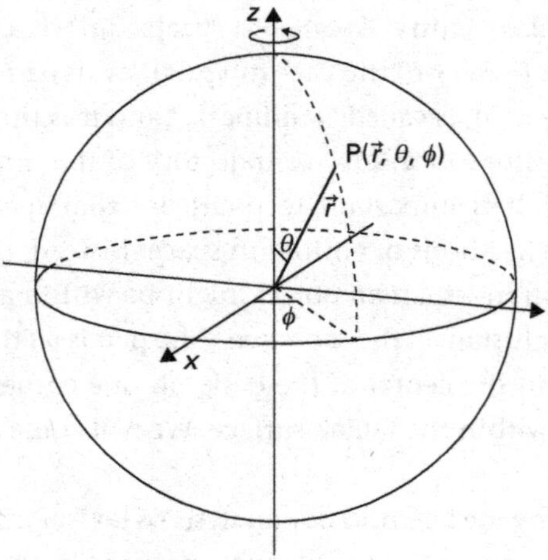

Figure 3.1
Polar coordinates.

just as with nouns and adjectives, spatial prepositions correspond to convex regions in conceptual spaces.

Now, following Zwartz and Gärdenfors (2016), we will take the center of mass of the landmark (*L*)—a point—in a spatial sentence as the pole of the coordinate system.[7] We will also assume that landmarks are spherical in three dimensions or circular in two, with radius *r*. Then, the preposition in a spatial sentence will determine a region of the space in which the trajector, represented as an arbitrary point, may lie.

Let us illustrate this by analyzing the conceptual space of the preposition "*CloseTo(L)*." Because "close-to" only constrains *distance*, not direction, we can reduce the dimensions and represent it in a horizontal plane with ordinary polar coordinates $\langle r, \theta\rangle$. Thus, a trajector *T* is close to a landmark *L* *iff* $T \in \{\langle r, \theta\rangle: c < r < r_L\}$ (see figure 3.2). This region will be convex according to the definition of polar convexity. Notice that *c* has to be contextually defined. For instance, the *c* in the sentence "Munich is close to Augsburg" will be much larger than the *c* in "Maria is sitting close to Juan."

Lexical inferences occurring between prepositions that are polar antonyms (e.g., *far:close*, *in:out*, and *over:under*) can be explicated in similar terms. In general, for a pair of polar antonyms *x:y*, the negation of one of them implies the affirmation of the other. The proposition "*T* is close to

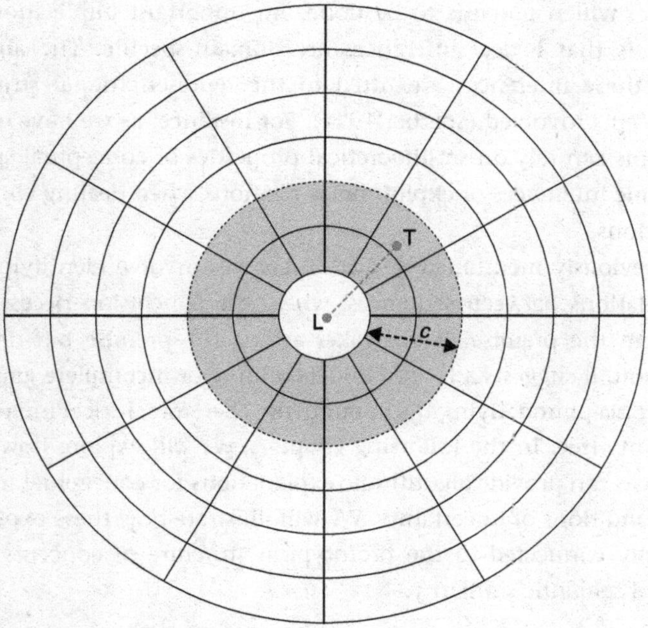

Figure 3.2
The regions representing *CloseTo(L)*.

L" entails "*T* is not far from *L*" since their representations are equivalent: $T \in \{\langle r, \theta \rangle: c < r < r_L\}$ *iff* $T \notin \{\langle r, \theta \rangle: r > c\}$. Again, inferring one from the other is explained as the reprofiling of the landmark and the trajector within a common conceptual base.

Here, we have only given one example of how to analyze the meaning of prepositions with the aid of polar convexity in conceptual spaces. Zwarts and Gärdenfors (2016) show that similar analyses can be given for most locational and directional prepositions.

3.3 Conclusion

The purpose of this chapter has been to present, through some examples, a strategy to build algorithmic explanations of lexical inferences in terms of conceptual spaces. Nevertheless, a systematic model of such inferences would require generalizing this kind of analysis to account for the inferential patterns associated with the most common lexical types in natural language, which remains to be done. An important implication of our analysis is that lexical inferences are domain-specific: The algorithms driving these inferences are fitted to the representational structure of the concepts involved (Sperber 1994). For instance, as we have observed, algorithms can rely on set-theoretical properties of conceptual spaces for taxonomic inferences or exploit polar relations when dealing with spatial prepositions.

As previously mentioned, lexical inferences involve identifying entailment relations between sentences, where the conclusion necessarily follows from the premise. If a speaker asserts the premise but denies the conclusion, it suggests a lack of understanding or incomplete grasp of the lexical concept underlying the entailment. Therefore, lexical inferences are uncertainty-free. In the following chapters, we will explore how conceptual spaces can provide algorithmic explanations for conceptual inferences under conditions of uncertainty. We will illustrate how these explanations are deeply connected to the prototypical structure of concepts and the notion of semantic similarity.

4 Induction as Knowledge-What

4.1 The Problem of Induction

In the previous chapter, we saw a class of inferences that are truth-preserving and rely on core lexical knowledge.[1] Yet, much of our everyday reasoning does not fit this pattern. Instead, it is *inductive—ampliative* in nature—and, as such, involves varying degrees of uncertainty as the conclusions extend beyond the information in the premises. Inductive reasoning is one of the most impressive aspects of human cognition.[2] We tend to generalize from a limited number of observations, sometimes with overwhelming confidence, and often with success. A key challenge in philosophy and cognitive science is to understand and model the mechanisms underlying inductive reasoning.

We do not perform inductive inferences in an arbitrary manner. Peirce notes that there are certain forms of constraints that delimit the vast class of possible inferences:

> Nature is a far vaster and less clearly arranged repertory of facts than a census report; and if men had not come to it with special aptitudes for guessing right, it may well be doubted whether in the ten or twenty thousand years that they may have existed their greatest mind would have attained the amount of knowledge which is actually possessed by the lowest idiot. But, in point of fact, not man merely, but all animals derive by inheritance (presumably by natural selection) two classes of ideas which adapt them to their environment. In the first place, they all have from birth some notions, however crude and concrete, of force, matter, space, and time; and, in the next place, they have some notion of what sort of objects their fellow-beings are, and how they will act on given occasions. (Peirce 1955, 214–215)

Here, Peirce hints at an *evolutionary* explanation of the effectiveness of inductive reasoning, claiming that "the human intellect is peculiarly adapted to the comprehension of the laws and facts of nature" (1955, 213).

Inductive inference was central to the logical positivists, playing a key role in their verificationist program (Carnap 1950; Hempel 1965; Creath 2014; Vickers 2014). Following the standard analysis of deductive inference, the positivists—and later the broader analytic tradition—conceived of induction as a specific (and formalizable) relation between sets of sentences.

However, it soon became apparent that the logical approach led to paradoxes. The most well known are Hempel's (1965) "paradox of confirmation" and Goodman's (1955) "new riddle of induction." Restricting the analysis of induction to the logical form of sentences forces us to treat all predicates on a par, generating symmetries that conflict with our intuitions about permissible inductive inferences: "Raven" in Hempel's paradox is treated on a par with "non-raven," "green" in Goodman's with artificial predicates like "grue." What is required is a nonlogical way of distinguishing the predicates that may be used in inductive inferences from those that may not.

Goodman (1955) tries to solve the problem by distinguishing between "entrenched" concepts such as "blue" and "green," which have a successful history of usage within our inferential practice, and concepts such as "grue" and "bleen," which do not. Goodman's distinction, however, provides little more than a name of the problem. He does not give any solution to the problem of how to determine whether a concept is entrenched or not.

Our approach diverges from the traditional propositional or sentential framework, embracing a *cognitivist* and *naturalistic* perspective. We posit that inductive inferences arise from our cognitive faculties (Quine 1969; Humphrey 1992; Kornblith 1993; Gärdenfors 1990, 2000), shaped by evolution to discover, recognize, and categorize certain patterns in the world—otherwise, our generalizations and predictions would be miraculous (Dennett 1991; see also Johansson 1998). Our first topic concerning inductive inferences will be the kind of *knowledge* underpinning them.

Our diagnosis of why the paradoxes have emerged in the traditional treatment is that induction does not primarily concern relations between sentences but between *concepts*. We outline a new approach to induction that is based on three theses. The first thesis says that it is possible to distinguish between *natural* and *nonnatural* concepts and that only natural concepts occur in inductive inferences. The second thesis introduces *knowledge-what*: a distinct epistemological category that captures the interplay between properties and concepts, irreducible to mere *knowledge-that* or *knowledge-how*. The third thesis is that the role of induction is to generate knowledge-what. This entails that we find much of the earlier discussion of induction misguided since it has focused on induction as generalizations generating knowledge-that.

In this context, it should be noted that there are two meanings of "generalization" in the literature. One is logical, relating to the connection between sentences describing individual instances and universal sentences covering the individual sentences. The other, also called "stimulus generalization," is psychological and concerns the relations between reactions to a particular stimulus and a class of *similar* stimuli. We argue that human inductive inferences are more related to the psychological notion of generalization.

4.2 Natural Concepts

The literature offers various proposals for determining which concepts can legitimately feature in inductive inference. One influential approach distinguishes between concepts that denote "natural kinds" (or "natural properties") and those that do not, restricting inductive reasoning to the former. Natural kinds are normally interpreted realistically, following the Aristotelian tradition (Hacking 1991), and therefore assumed to represent something that exists in reality independently of human cognition. Yet, for understanding inductive inference, the mere existence of such properties is insufficient; we must also account for how humans grasp and employ natural kinds in reasoning. In other words, what is required to understand induction, as performed by humans, is a cognitive analysis of natural concepts.

Several criteria for what constitute natural concepts have been proposed. As mentioned in section 2.4, the criterion that the regions of a conceptual space representing a concept be *convex* is central. This criterion has the following consequences (see Gärdenfors 2024b):

(4.1) *Prototypicality*: It explains why concepts show prototype effects. Points in a convex region are more or less central, and central points are prototypical.

(4.2) *Generalizability*: Convexity provides a model of how concepts can support generalizations. If two objects have been observed to fall under a concept, then, by convexity, all objects located in between in a conceptual space also fall under the concept.

(4.3) *Learnability*: Convexity makes learning more efficient. By generalization, a few observations of instances of a concept provide information about the region of a conceptual space that is covered by the concept.

(4.4) *Communication*: Convex concepts make it easier for people to use the concept in the same way (Jäger 2007; Warglien and Gärdenfors 2013).

Further criteria for natural concepts have been proposed by Regier et al. (2007) and Douven and Gärdenfors (2019):

(4.5) *Parsimony*: The conceptual structure should not overload the system's memory.

(4.6) *Informativeness*: The concepts should be informative, meaning that they should jointly offer good and roughly equal coverage of the domain of classification cases.

(4.7) *Representation*: The conceptual structure should be such that it allows the system to choose for each concept a prototype that is a good representative of all items falling under the concept.

(4.8) *Contrast*: The conceptual structure should be such that prototypes of different concepts can be so chosen that they are easy to tell apart.

(4.9) *Well-formedness*: The concepts should be "well-formed" in that the items falling under any one of them are maximally similar to each other and maximally dissimilar to the items falling under the other concepts represented in the same space.

(4.10) *Coherence*: The features associated with a particular concept are perceived as "belonging together" (Osta-Vélez and Gärdenfors 2025).

These criteria should be seen as desiderata for natural concepts. *Coherence* will be discussed in chapter 9. It may be difficult to satisfy all of them. For example, to some extent, *informativeness* clashes with *parsimony*, since the more concepts are covering a conceptual space, the more informative they are. So, the problem is to strike the right balance between *informativeness* and *parsimony*. In relation to the vertical dimension of prototype theory (section 2.5.2), it has been noticed that the step from a superordinate level to the basic level typically involves a large increase in informativeness, while a step from the basic level to a subordinate level typically does not markedly increase the information content of the concepts. This can be seen as an argument for selecting the basic level as the most natural level of concept representation.

4.3 Knowledge-That, Knowledge-How, and Knowledge-What

4.3.1 Induction from the Perspective of Language

Historically, the "empiricist turn" of the seventeenth century raised interest in inductive inferences, although it remained uncertain how induction should be justified since it lacked the logical rigor of deduction. Hume (1748/1999) argued that it is impossible to justify inductive inferences, although he acknowledged *habit* as an inevitable part of human reasoning. Other issues included pinpointing which evidence, and what amount, was enough for valid inductive inferences and finding methods that could separate good inferences from bad ones (Mill 1843; Vickers 2014).

In the mainstream debate within analytic philosophy, a major distinction has been between *knowledge-how* and *knowledge-that* (Ryle 1949). It has been a tacit assumption that induction does not concern knowledge-how. As part of the linguistic turn of analytic philosophy, there was a preference for analyzing inferences, including induction, as relations between sentences. Hence, it was concluded that if induction is an epistemic process, it must deal with knowledge-that, since knowledge-that is propositional and can be expressed in sentences.

For the logical positivists, the basic objects of study were sentences in some more or less regimented language. Ideally, the language was a version of first-order logic where the atomic predicates represent observational properties. These observational predicates were taken as primitive, unanalyzable notions. A consequence of this methodology was that all observational predicates were treated similarly since there were no logical reasons to differentiate between them (see section 1.2).

In this tradition, particular observations are used as evidence for inductive generalizations (Carnap 1950, 1971; Hempel 1965). When connections are found within the registered observations, inductive generalizations can be made, which can be confirmed by additional observations. So, if observations of objects o_1, o_2, o_3 . . . falling under a concept *O* all are *C*, the generalization that *all O* are *C* can be made. The inference thus concerns a relation between individual and universal sentences. The evidence from the premises gives stronger or weaker support for the conclusion. Since this inductive process does not have the same logical rigor as a deductive process, the methodology of induction requires that supporting evidence preferably (1) should come in large numbers, (2) come from several different contexts, and (3) have no negative cases (Hempel 1965).

One point that has been downplayed in the debate, however, is that not all universal sentences can function as conclusions in inductive inferences. Since Aristotle's classic "All men are mortal," inductive inferences only involve universal sentences that are *generics*—that is, express relations between kind concepts and properties. An accidental universal sentence such as "All persons in this room are Swedish" would not be acceptable as an inductive inference, even when perfectly supported by the given evidence. Such "accidental generalizations" do not support counterfactuals of the form "If a person came into the room, he or she would be a Swede." So, even though the logical form of a law-like sentence is the same as that of an accidental universal sentence, we submit that law-like sentences function as generics (see chapter 6).[3]

4.3.2 Knowledge-What as Knowledge of Concepts

Ryle (1949) provides some influential arguments for upholding the distinction between "knowledge-how" and "knowledge-that." He argues that knowledge-that is to *possess* knowledge, whereas knowledge-how is to be *intelligent*. Knowledge-that thus concerns relations between agents and true propositions, whereas knowledge-how instead concerns abilities, dispositions, and actions of the agent.

Fantl (2016) proposes an extension of the traditional dichotomy between knowledge-how and knowledge-that: "There's the kind of knowledge you have when it is truly said of you that you know a person—say, your best friend." In our opinion, Fantl's knowledge of "acquaintance" is a special case of a third type of knowledge. We want to single out the ability to *identify kinds*—in particular, to know *the relation between kind concepts and properties* as a special form of knowledge, which we call *knowledge-what*.[4]

4.3.3 Knowledge-What Is Separate from Knowledge-That

Even though knowledge-what is primarily nonlinguistic, it can be expressed in language. We offer two arguments for distinguishing knowledge-what from knowledge-that, which apply even when both are linguistically formulated. The first one builds on the observation that it seems perfectly natural that the following two sentences can be accepted simultaneously:

(4.11) Spiders have eight legs.

(4.12) The spider in the kitchen window has only seven legs.

However, if (4.11) is expressed—as is standard in writings on induction—in the form of a universal sentence:

(4.11′) All spiders have eight legs.

Then (4.11′) contradicts (4.12). There are two reasonable ways out of the contradiction:

(1) Deny that the creature in the kitchen window is a spider.
(2) Deny that (4.11) expresses the same knowledge as (4.11′).

The case against option (1) lies in conceptualizing a spider's "essence" as a "pattern" of properties. A seven-legged spider retains its kind-identity through other "essential" or definitional traits.[5] In favor of option (2), it is worth highlighting that (4.11) expresses definitional properties, while (4.12) merely expresses an accidental fact. Interpreting (4.11) as (4.11′) and putting it together with (4.12) conflates the two different types of knowledge. As an alternative way out of the contradiction, one may propose the following formulation of (4.11):

(4.11″) Spiders *characteristically* have eight legs.

Barring the problem of explaining the meaning of "characteristically" in a noncircular way, we consider that this formulation supports our position that the knowledge expressed in (4.11) is of the definitional or characteristic form, that is, knowledge-what.

Our take on knowledge-what is that it concerns defining and characteristic knowledge, while knowledge-that concerns facts—accidental facts as well as facts of the type 2+2=4. Our central thesis, to be discussed in section 4.6, is that inductive inferences, as a subset of knowledge-what, result in knowledge about characteristic properties. We thus heed the distinction between knowledge-how and knowledge-that while adding knowledge-what as a third type of knowledge, which is central for processes of induction.

4.4 Memory and Different Forms of Knowledge

Without memory, there is no knowledge. In this section, we take a cognitive and neuroscientific perspective and present a different kind of support for our thesis that knowledge-what is a separate form of knowledge by mapping our partitioning of the three types of knowledge onto Tulving's (1985) partitioning of long-term memory into procedural, semantic, and episodic.

The nondeclarative *procedural* memory handles an agent's skill in performing a task. This kind of memory can be described as generated by an automatic process, where an agent learns and remembers *how* to do something. Learning is achieved through repetition or practice, and procedural memory can easily be associated with operant conditioning since it is possible to describe in terms of stimulus and response. Procedural memory is something humans share with many other animals (Tulving 2002).

Semantic memory allows agents to actively cognize properties and concepts (Herrnstein 1990; Martin et al. 1996; Martin and Chao 2001; Binder and Desai 2011; Yee et al. 2014). Semantic memory is general and does not depend on specific references to experiences. This kind of memory is needed for handling the environment as efficiently as possible. In particular, semantic memory is crucial for mapping concepts to actions.

Episodic memory governs experienced knowledge that can be used in narratives. This kind of memory generates self-aware *remembrance* concerning single events such as they are experienced from a first-person perspective (Tulving 1985, 2002). Episodic memory makes it possible for humans to "time-travel" in their minds. It allows us to remember individual events or episodes and the order in which they have occurred. Tulving (2002) claims that this form of memory is only found in humans.

Tulving's three memory systems, while distinct, likely operate in parallel (Tulving 2002, 6; Yee et al. 2014). We propose a direct correspondence between these systems and our tripartite knowledge framework: Procedural memory handles knowledge-how, semantic memory handles knowledge-what, and episodic memory handles knowledge-that.[6] Since the characterization of the knowledge handled by the three memory systems clearly maps onto our description of the three kinds of knowledge, this mapping supports that the three types we distinguish indeed have different functions in human cognition.

Given the evidence from neuroscience, the credibility of our tripartite account of knowledge offers a counterargument against a reduction of knowledge-what (or knowledge-how) into knowledge-that. Conceptual knowledge-what should not be seen as reducible to propositional knowledge-that, since the memory system underlying knowledge-what is different from that underlying knowledge-that.

The upshot is that the mapping between types of knowledge and long-term memory systems provides us with a naturalistic argument for separating knowledge-what from knowledge-that. In brief, knowledge-what is a special form of knowledge, just as semantic memory is a special form of memory.

4.5 Knowledge-What as Relations Between Concepts and Properties

We might not have unbiased contact with the world, but it is still the real world that provides the sensory input we get. As Kornblith (1993, 35) argues, the possibility of inductive knowledge stems from the world having the required causal structure to enable the formation of natural kinds. This is because certain clusters of properties are organized in a sufficiently stable manner to cohere into natural categories, thereby providing the foundation for our inductive reasoning processes.

As a way of capturing clusters of properties, Criterion *C* introduces relations between domains as a factor of an object concept. We propose that knowledge-what consists of such relations. For example, knowing what aspartame is involves knowing the relation between the chemical domains that characterize aspartame and the sweetness region of the taste domain.

There are, however, different kinds of relations. The strongest one is when all examples of a category fall within one region of a domain. This case can be divided into definitional relations, such as in "All dogs are mammals" and empirical, such as in "All ravens are black." Another form of relation is *covariation*; for example, "Metals expand when heated" describes a covariation between the temperature and size domain. Another example is the covariation between the color and sweetness of fruits.

Even though there are many possible relations between categories and properties, we can often discern which relations are relevant. We have a built-in understanding of the world's structure, where some properties and patterns are intuitively grasped (Kornblith 1993, 100–101; Johansson

1998). As Kornblith (1993, 104) points out, "We are accomplished detectors of multiple, clustered patterns of covariation."[7]

An argument for focusing on covariations between domains comes from work by Billman (1983) and Billman and Knutson (1996), which indicates that humans are quite good at detecting covariations that cluster several domains (Hayes et al. 2010). A plausible explanation of this phenomenon is that our perception of natural objects shows covariations along multiple domains, and, as a result of natural selection, we have developed a competence to detect such clustered relations. In line with this, the basic-level categories of prototype theory (Rosch 1975) can be characterized by distinctive clusters of covariating properties (Holland et al. 1986, 183–184). And, as we will argue in chapter 9, covariation between several domains can be seen as a criterion of *coherent* concepts.

4.6 Induction as Generating Knowledge-What

As we mentioned earlier, the propositional approach to induction led to counterintuitive conclusions visible in numerous paradoxes. Quine's (1969) negative conclusions concerning the possibilities of defining "natural kind" or the corresponding notion "similarity" can be interpreted as indicating that we must go beyond language to find a solution. What is needed is a way of tapping our sources of knowledge so that we become able to distinguish the properties that may be used in inductive inferences from those that may not.

For Goodman (1955), the question of what makes certain generalizations law-like becomes the problem of which predicates are "projectable"—that is, which predicates can be used in inductive inferences.[8] The solution we propose here is that only natural properties and categories, as defined in criteria *P* and *C* in section 2.4, are "projectable"—that is, allowed in inductive inferences (Gärdenfors 1990, 2000; Gärdenfors and Stephens 2018). Consequently, the feature of a conceptual space most essential for a theory of induction is its topological and metric properties, while the logical structure of the language that "lives on" the conceptual space is secondary.

As an example, let us take a brief look at the concepts that occur in Hempel's (1965) paradox of confirmation. The paradox describes how all observations of black ravens confirm the generalization that all ravens are black. However, since "All ravens are black" is logically equivalent to "All non-black things are non-ravens," all observed non-black non-ravens logically confirm the same generalization. Observing a white shoe, for example, would confirm that all ravens are black. It seems odd that any such observation should support an inductive inference that all ravens are black.

According to Criterion *C*, kind concepts are convex regions in product spaces, where each subspace represents a property of the concept. If the properties of the concept all correspond to convex regions, then the product of the regions will be convex too. In contrast, the concept "non-raven" would be difficult to count as a natural concept. The class of all objects that are non-ravens are instantiated in many unrelated domains. The associated regions, let alone their product, cannot be specified as a convex region of some space. Consequently, "non-raven" does not qualify as a natural concept. A similar analysis can be provided for Goodman's (1955) example of "grue" (Gärdenfors 1990, 2000). The properties used in these problems do not correspond to convex regions in domains and are hence not projectable.

Conceptual spaces thus help us understand induction as a way of achieving knowledge-what. Given the characterization of projectable predicates as natural properties and categories, our analysis of what is achieved in induction is knowledge about relations between such properties and concepts. Induction lets us achieve new knowledge-what about concepts. To know what properties a specific category is related to is to have knowledge about characteristic properties of that concept. Most scientific empirical discoveries are of this type. For example, when it was discovered that penicillin is an antibiotic, such a relation was established (Aldridge et al. 1999; Lax 2004). Or when it was discovered that a certain alloy of niobium and titanium was a superconductor (Berlincourt and Hake 1963), new knowledge about characteristic properties of the type knowledge-what was acquired.

Induction consists of generalizing from a limited number of observations. In logical approaches, "generalization" means forming some form of universal sentence. However, the situation is different when conceptual spaces are used as a basis. The similarity structure of the domains allows generalization in the form of extending the given observations to *similar* instances, in particular by applying the convexity criteria *P* and *C*. This form of generalization, therefore, comes closer to what is called "stimulus generalization" in psychology. Notably, this form of generalization has not been discussed in philosophical literature on induction.

It should be noted that the change in how induction is perceived—from relations between sentences to relations between properties and kind concepts—does not lead to any radical changes in the *methodology* used to establish inductive knowledge. Well-known requirements of repeated experiments, precision, variation, and generalizability in experiments are still valid (Hempel 1965; Seltman 2015). Furthermore, these requirements turn out to be even more natural from the perspective of establishing knowledge-what. As remarked above, generalizability achieves a different meaning, and different methods for determining relations that take distances in domains into consideration should therefore be put in focus.

It is time to give up the focus on knowledge-that in analytical philosophy. In our opinion, many of the riddles of induction are a consequence of this focus, and they will not appear if knowledge-what is accepted as a type of knowledge and induction is recognized as involving it. By introducing the tripartition of knowledge, we hope to reboot epistemology in a naturalistic direction. Since philosophical (but not psychological) research on inductive processes during the last century has focused on propositional knowledge-that, we propose that the representations of concepts and properties—as a way of modeling knowledge-what—should be given much more attention.

5 Reasoning Based on Expectations

In chapter 3, we showed how conceptual spaces can model entailment relations between sentences, thereby explicating our intuitions of the adequacy of lexical inferences. Although this type of inference is not strictly logical, it is akin to deductive inference in being *informationally conservative*: The content of the conclusion is contained within the conceptual representation of the premise. However, most of our inferential activity takes place under uncertainty, as the information in the premises is often incomplete and thus does not fully determine the content of the conclusion. In this chapter, we will show how conceptual spaces can model inference under uncertainty by considering it an expectation-based process.

It should be noted already from the beginning that by "uncertainty" we do not mean "probabilistic," as is common in decision theory. In contrast, we shall argue that probabilistic models may not be suitable when it comes to reasoning with concepts.

5.1 What Are Expectations?

We all have expectations. For instance, if we know that Maria is from Spain, we expect her to have dark hair and to speak Spanish; or if we are driving a car and we spot a person waiting at a pedestrian crossing, we expect her to cross it. In general, our expectations about the world are crucial for guiding our reasoning and actions in everyday life, and they build directly on the structure of our background knowledge.

As a matter of fact, expectations are ubiquitous, although they often operate implicitly. Though you probably do not consciously think about it, you expect there to be a floor when you enter a room; you expect a door handle not to break when you press it; you expect the train to arrive at its scheduled time; and you do not expect a cat to laugh. Central to this chapter is the idea that expectations, which are deeply intertwined with our conceptual system, shape our everyday reasoning in various ways. In brief, we use them as supplementary premises when we make inferences.

Our expectations, however, are not infallible. For instance, upon learning that something is a bird, one typically expects it to fly. Yet, upon discovering it is an emu, that expectation is overturned. This example illustrates our capacity to draw accurate conclusions even when our initial expectations are contradicted.

It has been argued that much of so-called nonmonotonic logic can be reduced to classical logic with the aid of an analysis of the expectations working as hidden premises in arguments (Gärdenfors 1994; Gärdenfors and Makinson 1994). The guiding idea is that when we evaluate whether a conclusion *Q* follows from a set of premises *P*, the background information that we use for inferring does not only contain the premises in *P* but also information about what we expect in the given situation so that we end up with a larger set of assumptions.

Expectations include not only our knowledge as a limiting case but also other beliefs regarded as plausible enough to be used as a basis for inference as long as they do not give rise to inconsistencies. Thus, expectations are *defeasible* in the sense that if the premises in *P* conflict with some of the expectations, we do not use them when determining whether *Q* follows from *P*.

It is important to distinguish between two types of expectations: those based on facts and those based on concepts. If the weather forecast predicts rain, I will form a *factual* expectation. If I learn that elephants are long-lived, the expectation I will form will be grounded on a relation between the object concept ELEPHANT and the property *long-lived*. Here, we focus on expectations based on conceptual relations—that is, on knowledge-what. Such relations are often expressed linguistically as generics. We return to the function of generics in chapter 6.

Expectations based on knowledge-what can be expressed as default assumptions—that is, statements about what is normal or typical. However, when evaluating their role in arguments, it is important to note that they do not all have the same strength. For example, in certain cases, we consider the relation among two concepts, such as "Tigers are striped," to be strong enough to work as a rule that is almost universally valid so that an exception to it would be extremely unexpected, while in other situations this relation could be better described as rules of thumb that we use for want of more precise information. For instance, while walking on the sidewalk, we expect the ground to be solid enough to support our body weight, but when we are hiking in the snow, this expectation will be weaker, and therefore we will walk carefully to avoid sinking. An exception to the latter type of rule is not unexpected to the same degree as in the former case. In brief, our expectations are all defeasible, but they exhibit varying degrees of defeasibility.

In this chapter, our main objects of study are expectation orderings as a way of expressing degrees of defeasibility. An argument in favor of using expectation orderings for explaining everyday reasoning is that they contain enough information to express, in a very simple way, what we require with respect to default information. The principal idea is that a default statement of the type "*F*s are normally *G*s" can be expressed by saying that "if something is an *F* then it is more expected that it is *G* than that it is non-*G*." This formulation is representable in an expectation ordering $>$ by assuming that the relation $F \rightarrow G > F \rightarrow \neg G$ holds (Gärdenfors and Makinson 1994).

Our position is that a proper explication of the role of expectations in reasoning must rely on a model of the structure of *background knowledge*. Although this notion has played a central role in several areas of philosophy and logic over recent decades, few efforts have been made to properly define it. These disciplines have generally adhered to the formalist view, assuming that implicit and explicit knowledge could be represented as propositional knowledge-that within a cognitive agent's "belief box."

Within computer science, new representational frameworks were developed for organizing knowledge structures more efficiently. These included frames (Minsky 1975), semantic networks (Quillian 1967), and conceptual graphs (Sowa 1991), all non-propositional alternatives to classical logic that presented new ways of representing information. However, these methods often lacked psychologically realistic foundations and did not explain the origin of their semantic organization (Brachman 1979; Moore 1982).

Artificial neural networks, particularly those used in deep learning, also offer a powerful alternative to symbolic approaches, with their ability to learn from vast amounts of data and capture complex patterns. However, they too have faced criticism for their opacity and the difficulty in interpreting the learned representations (Marcus 2018).

This chapter develops a cognitively grounded theory of expectation orderings through conceptual spaces. While Gärdenfors and Makinson (1994) provided a formal analysis of the role of these orderings in reasoning, they left their cognitive origins unexplored. We address this gap by showing how concept structure shapes expectations.

Our analysis unfolds in several steps. In section 5.2, drawing on a Gricean principle, we examine how the information a person has about a concept's structure influences her expectations. Section 5.3 uses the theory of conceptual spaces to show how degrees of defeasibility are expressible in terms of prototypes and distances. Section 5.4 shows that prototypes can also be used to update expectation orderings, while section 5.5 argues that the default rules that have been used in many areas of nonmonotonic logic can be derived from the expectation orderings. Section 5.6 presents an explication of some alleged reasoning "fallacies" that use expectations rather than probabilities. Section 5.7 discusses inferential strength, and 5.8 finally describes the meaning of "but" in terms of expectations.

5.2 A Gricean Principle

As mentioned earlier, our expectations about the world are intrinsically linked to the structure of our background knowledge. For example, if we learn that Maria adopted a new pet, we would likely expect it to be a typical choice, such as a dog or cat or perhaps a bird or fish. We might be surprised to discover that Maria's new pet is a lizard, as these animals are less expected to be domestic pets. However, our expectations can shift based on context. In regions where lizards are common pets, learning about Maria's new lizard would be less surprising. Expectations, just as typicality, are strongly dependent on cultural contexts (Schwanenflugel and Rey 1986; Lin et al. 1990).

This example shows two important features of expectations: (1) it is a graded and subjective phenomenon, and (2) the strength of an expectation depends, to a significant extent, on the prototypical structure of concepts. To be more precise about the second point, when we receive information about an object belonging to a concept, we tend to reduce subjective uncertainty by implicitly assuming the object possesses other typical properties associated with that concept. The effects of this will depend on the structure of that concept and its semantic associates. As we will show, conceptual spaces can give us useful tools for articulating such relations.

As we said before, the main idea is that when we categorize an object x as C, we implicitly form expectations about properties that this object is supposed to have because of falling under C. Depending on the strength of the expectations, the properties can be ordered, thus generating an expectation ordering. The organization of this ordering will depend upon the prototypical structure of C, which can be represented in its corresponding conceptual space.

The underlying rationale for this method of generating an expectation ordering is a version of the Gricean *maxim of quantity* (Grice 1975), which states that, for the sake of efficient communication, a speaker should be as informative as required for the purposes of the exchange, neither more nor less.[1] If you are informed that an object x should be categorized as, for example, a bird, but you do not know more about what kind of bird x is, then you will expect that x has all the prototypical properties of birds: You expect that x has wings, that it has a beak, that it builds nests, that it sings, that it flies, and so on. In other words, you assume these properties unless informed otherwise, since the maxim of quantity implies that the speaker would have specified a more unusual kind of bird if those expectations were not met.

Furthermore, when new information is added, expectations are restructured. If, after learning that x is a bird, you learn that it is an ostrich, you will no longer expect that it flies, or that it sings. Instead, some new expectations will be added, such that x is big, that it runs fast, and that it kicks hard.[2] Understanding how expectations are generated and organized when some information is received and how they are restructured when new information is added are the two main aspects that a model of this phenomenon must account for.

5.3 Degrees of Defeasibility

As before, we use $C(M)$ to denote the representation of a concept M in a conceptual space. A sentence $M(x)$ expresses that an object x falls under M. Our main claim is that the expectations toward $M(x)$ are structured around the prototype p_M of M, which is also an n-dimensional point

$p_M = \langle p_{M_1}, p_{M_2}, \ldots, p_{M_n} \rangle \in C(M)$. In other words, if the only thing we know about x is that it falls under concept M, we will expect it to be represented as (close to) p_M—that is, to have all the properties of the prototype.[3]

Now, the expectations elicited by the sentence $M(x)$ may well go beyond the specific properties determined by the prototype. They extend to all the possible properties that an object falling under M may have. In the conceptual space framework, this means that representing an object under a concept M implies that it may occupy any possible position in the region $C(M)$. Different positions imply different properties for the object. The properties that do not apply to p_M can be considered secondary expectations since they are weaker (more defeasible) than the ones that apply to p_M. We claim that for any property R_i in $C(M)$, its *degree of defeasibility* can, in general, be specified as a function of its position relative to the prototype of M. The challenge now is whether it is possible to construct an ordering of properties that reflects their "degree of expectedness" (and thus, their degree of defeasibility).

Our typicality measure T_M (section 2.7.2) applies to every property in a conceptual space and, consequently, can be used as a basis for constructing the desired ordering of properties. We define the set $Exp(M) = \{R_1 > R_2 > \ldots > R_M\}$ as the expectation ordering of properties associated with a concept M as follows: Given two properties R_i, R_k in $C(M)$, R_i is more expected than R_k—written $R_i > R_k$ if, and only if, $T_M(R_i) > T_M(R_k)$. Recall that according to the typicality measure, all prototypical properties are more typical than all non-prototypical properties. This ordering of properties can be turned into an ordering of atomic sentences by saying that for all individuals x, $R_i(x) > R_k(x)$ if, and only if, $R_i > R_k$.[4]

Let us consider an example using the fruit space introduced in section 2.4, with dimensions of color, taste, shape, and texture. Imagine we are told that a is an apple. Our primary expectations would be that a possesses the properties of a prototypical apple: *round*, *sweet*, *red*, and *smooth*. However, despite all being fulfilled by the prototype, these properties vary in their degree of typicality. For instance, sweetness is more typical of apples than redness. We would be more surprised by a non-sweet apple than a non-red one. Consequently, in our conceptual space, points representing non-red apples would be closer to the prototype than those representing sour or bitter apples. Similarly, bitterness is an unusual property for

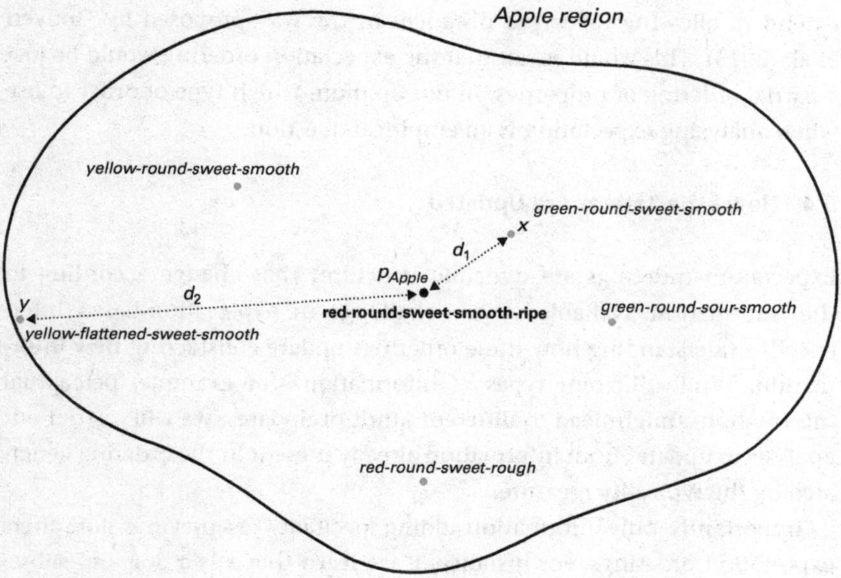

Figure 5.1
An illustration of an apple region of the fruit space, with points representing more and less typical instances of apples.

apples, certainly less expected than being yellow. Thus, instances of yellow apples will be closer than the prototype than instances of bitter apples (see figure 5.1).

An expectation ordering of properties for APPLE can thus look like this: *Exp*(APPLE) = {*round* > *smooth* > *red* > *sweet* > *green* > . . .> *yellow* > . . .> *bitter* > . . .}.

The typicality measure T_M produces a fine-grained order of expectations, enabling comparisons between individual properties. This addresses the problem of determining the origin of expectation ordering mentioned earlier.

It should be noted that since it is based on a distance function, T_M generates a total ordering of individual properties for a given conceptual space. In many cases, however, this assumption is cognitively unrealistic. For example, which is the most typical property of an ostrich: That it runs fast or that it kicks hard? It would, therefore, be more natural to abstain from judging which property is more typical. In cases like this, the assumption of a unique distance measure on the conceptual space is too strong. One could extend it, allowing for vague distances in the way proposed by Douven et al. (2013). This would mean that the expectation ordering would be just a partial ordering of properties. In our opinion, which type of order to use when analyzing expectation is an empirical question.

5.4 How Expectations Are Updated

Expectation orderings are dynamic structures that change according to the information available. A key challenge in expectation-based inference is understanding how these orderings update considering new information. While different types of information—for example, perceptual information—might lead to different kinds of updates, we will restrict our analysis to updates from information already present in the ordering generated by the typicality measure.

Importantly, only information adding specificity to a previous state alters expectation orderings. For instance, if we learn that *x* is a dog and subsequently that *x* is a mammal, the expectation ordering remains unchanged, as the latter information is implicit in the initial state. According to the framework defended in this book, reasoning about object properties under uncertainty amounts to specifying the position of a point in an *n*-dimensional space. Each new piece of information with a more specific value in some of the dimensions reduces the space of uncertainty.

5.4.1 Prototype Shifts

Let us start by analyzing a kind of updating we name *prototype shift*. Consider an expectation ordering *Exp*(*M*) structured around prototype p_M. If we learn that an entity *x* is also *G* and we know that *G* is a subordinate concept of *M*—that is, $G \subset M$—then *Exp*(*M*) updates to *Exp*(*G*). This shift changes the prototype to p_G and restructures the ordering accordingly (following the typicality measure). This will typically reduce the number of properties in the original expectation ordering since the new information will "shrink" the relevant area of the conceptual space.

For example, suppose that we are told that *x* is a bird. An expectation set will be generated from p_{BIRD}. *Exp*(BIRD) encompasses numerous color properties and shapes, reflecting the wide variability of the bird category. If we then learn that *x* is a penguin, the expectation ordering updates to *Exp*(BIRD&PENGUIN). This shifts the prototype to that of penguins, eliminating most of the elements referring to colors that are not black and white, as well as the elements referring to not penguin-like shapes. Also, it will lose the property of flying in the domain encoding information about locomotion.

5.4.2 Updating via Properties

A subtler kind of update occurs when information about new (compatible) properties is added. Given the information *M(x)* and its associated ordering *Exp*(*M*), if we learn that *F*(*x*), where *F* is a property (subregion) of the domain $D_i \subseteq C(M)$, then *Exp*(*M*&*F*) will be equal *to Exp*(*M*) minus all the properties that are incompatible with *F* in D_i. If D_i is partitioned into disjoint regions, then *Exp*(*M*&*F*) will not include all the other properties of D_i that do not overlap with *F*. If this is not the case, then *Exp*(*M*&*F*) may or may not lose properties from *Exp*(*M*), depending on the structure of D_i.

Let us examine a simple example: We start with the information *Man(x)* and later learn that *Bachelor*(*x*), *Exp*(MAN) will shrink, losing the property *married*. However, if we subsequently learn that *a* plays golf as a hobby, we cannot eliminate other hobby-related properties from *Exp*(MAN&BACHELOR), since hobbies are not disjoint regions. A person might have multiple hobbies. Thus, for properties in not-disjoint domains, updates depend on the specific structure of the domain in question.

5.4.3 Updating via Correlated Properties

A more intricate update occurs when the new property correlates with another in a different domain. In the fruit space, for example, properties in dimensions like *ripeness* and *texture*, or *color* and *taste*, covary. Generally, apples are expected to be sweet, but green apples are typically sour. For such cases, the updating procedure will be the following.

Given *Exp*(*M*) and two correlated properties *G* and *F*, from domains D_i and D_k in *C*(*M*), when new information *G* is added, we will have that:

1. *Exp*(*M*) will be updated via property *G* as previously explained.
2. *Exp*(*M*&*G*) will have *F* as maximal or close to the maximal.
3. For any property *H* in D_k that is not correlated with *G*, *F* will be more expected than *H* in *Exp*(*M*&*G*).

Continuing with the apple example, suppose that our initial set is *Exp*(APPLE) = {*round* > *red* > *sweet* > *soft* > *green* > . . .> *yellow* > . . .> *bitter* > . . .}. If we are then told that apple *a* is green, properties like red and yellow, will be deleted, while the expectation of sour will become stronger and be more expected than any other taste-related property. The updated set would look like this: *Exp*(APPLE&GREEN) = {*green* > *round* > *sour* > *soft* > *sweet* > . . .> *bitter* > . . .}.

5.5 Degrees of Defeasibility

The limitations of classical logic in modeling reasoning under uncertainty have long been recognized by logicians and computer scientists. Different logical systems that incorporate rules with conceptual content have been developed to address these shortcomings. Among these, default logic (Brewka et al. 2007; Delgrande 2011; Horty 2012) stands out as a particularly influential approach. The central idea of this logic is to enrich the deductive system with default rules: predicate-specific inference rules meant to capture facts that are normally or typically the case when something is claimed to fall under some predicate *M*. For instance, the fact that most mammals have fur can be captured by the default $Mammal(x) \rightarrow HaveFur(x)$. Defaults like this can be used as rules of inference in nonmonotonic logical systems to extend derivations.

A foundational problem of default logic concerns the interpretation of the notion of a default rule (Delgrande 2011). Two epistemological issues affect this notion: (1) it is not clear where defaults come from, and (2) when multiple defaults can be applied to the same object, a decision method must be applied to determine which defaults have priority over the others (see Horty 2012, 19). A well-known example is the "Nixon diamond": By default, quakers are pacifists and, by default, republicans are not pacifists, but Richard Nixon is both a quaker and a republican. Thus, when determining whether Nixon is a pacifist or not, one must violate one of the default rules. Gärdenfors and Makinson (1994, section 3.3) argue that expectation orderings offer a natural way of representing defaults. In what follows, we will adapt and develop this idea for orderings based on conceptual spaces.

A natural answer to (1) is that defaults express strong regularities about phenomena that we can use as "inference tickets" in reasoning. However, from a cognitive modeling perspective, an internalist interpretation seems more appropriate: Defaults represent close conceptual relationships encoded in our background knowledge. This shifts the question of default origins to how they emerge from the structure of this background knowledge.

As Gärdenfors (2000, 117) suggests, the framework of conceptual spaces offers an answer to this question. Again, when we categorize an object under a concept, we automatically represent it as having a cluster of properties associated with the prototype of the concept. For instance, categorizing something as a fruit implies also categorizing it as sweet. This feature about categorization can also explain the kind of conceptual transitions expressed by default rules. An answer to the problem about the origins of defaults can be found, then, in the explanation that the theory of conceptual spaces offers about the prototypical structure of concepts and its exploitation in inference and categorization.

Let us now turn to (2). As we noted earlier, a default rule of the form "*Fs* are normally *Gs*" can be expressed in terms of expectations as "If something is an *F* then it is less expected that it is non-*G* than that it is a *G*." This can be represented in an expectation ordering as $F(b) \rightarrow G(b) > F(b) \rightarrow \neg G(b)$ for all b.[5]

Furthermore, the typicality measure gives us information about the comparative strength of each of these defaults. For instance, for the concept bird, it is to be expected that (according to the typicality measure) the property "build nest" has priority over "fly." This means that the default rule $Bird(x) \rightarrow Fly(x)$ will be weaker than $Bird(x) \rightarrow BuildNest(x)$. Generally, the earlier a property appears in the expectation ordering, the stronger its corresponding default rule will be.

The discussion above shows that the framework introduced in this chapter provides new insights for the two fundamental epistemological problems of default logic while suggesting a constructive solution to them.

5.6 Reasoning "Fallacies" and Typicality

5.6.1 Rationality and the Representativeness Heuristics

Cognitive psychology has historically operated under a "rationality assumption" (see Shafir and LeBoeuf 2002). As far as the mechanisms behind how beliefs are accepted are concerned, this rationality assumption has been operationalized through mathematical theories that provide normative frameworks for understanding concepts that we believe are key to reasoning: logic and probability theory.

In chapter 1, we saw some of the problems associated with understanding reasoning as a logico-formal process. Similarly, probability theory, the other pillar of theoretical reasoning, faced significant challenges due to the groundbreaking work of Kahneman and Tversky (1972, 1973). Their research revealed that human judgments under uncertainty systematically deviate from normative probabilistic and set-theoretical principles. For instance, when thinking about coin flips with a fair coin that can land heads (H) or tails (T), people typically consider a sequence like HTHTHTT more probable than HHHHHHT or any other sequence that appears to be nonrandom (Tversky and Kahneman 1974). Likewise, people would judge the claim of "an earthquake in California causing a flood in which more than 1,000 people will drown" as more probable than "a flood somewhere in the United States in which more than 1,000 people will drown" (Tversky and Kahneman 1983).

Despite this diagnosis, Tversky and Kahneman did not conclude that people are incapable of "extensional" reasoning. Instead, they observed that individuals can demonstrate adequate statistical intuitions in contexts in which the relevance of probabilistic or set-theoretic principles is clear (Tversky and Kahneman 1983). The problem is that such scenarios are uncommon in everyday life. More often than not, people rely on intuitive reasoning, employing "mental shortcuts" and heuristics to reach conclusions more quickly and with less cognitive effort (Gigerenzer and Goldstein 1996).

Among the many heuristics proposed by Kahneman and Tversky, one that stands out is the *representativeness heuristic*, according to which people swiftly judge the likelihood of something being the case based on how similar it is to a prototype stored in memory. While this inferential strategy is usually effective, it can also lead to errors since it downplays factors like sample sizes, prior odds, and some basic probability axioms, which are important for accurate likelihood estimation (Kahneman and Tversky 1972, 1973; Tversky and Kahneman 1983). Despite its explanatory appeal, the representativeness heuristic has been criticised for lacking a precise definition, an issue that would allow it to be fitted post hoc to almost any experimental result and would thus make it difficult to falsify (Gigerenzer 1996).

We contend that Kahneman and Tversky were fundamentally correct, and that several phenomena often labeled as reasoning fallacies are better understood as by-products of a representativeness heuristic, one that underlies concept-based reasoning and aligns with what we have called expectation-based reasoning. In the following sections, we will demonstrate how our theoretical framework can help specify the mechanisms underlying this heuristic, addressing Gigerenzer's critique.

5.6.2 Inclusion Fallacies

The first phenomenon that we will analyze was studied by Shafir et al. (1990) under the name of "inclusion fallacy," although later researchers renamed it as "modifier effect" (Connolly et al. 2007). In brief, an inclusion fallacy occurs when people judge a generic statement as less likely to be true when a *modifier* is added to the subject noun. For example, people judge "Lawyers are conservative" as more likely than "Young lawyers are conservative," even though the set of young lawyers is included within the set of all lawyers. Similarly, people judged the statement "Apples are red" as more probable than "Sour apples are red," even though sour apples are a subset of all apples.

Jönsson and Hampton (2012; see also Hampton et al. 2011) investigated this effect in depth and demonstrated that its strength was dependent on the typicality of the modifier: The less typical the modifier, the stronger the effect. For example, compared to the generic "Ducks have webbed feet," the modified generic "Baby ducks have webbed feet" (atypical modifier) was judged to be less likely than "Quacking ducks have webbed feet" (typical modifier). This occurs because atypically modified categories are perceived as less representative of the general category than typically modified ones, leading to greater uncertainty about whether they inherit the same attributes.

Our explanation of this phenomenon builds on a hypothesis about the nature of this kind of statement. While "common" generics convey prototypical information about a category—as will be discussed in chapter 6—modified generics aim at communicating information about property distributions that are typical of *meaningful subclasses* of the modified concept. As in the formation of expectation orderings, Grice's maxim of quantity plays a role here: Speakers must provide just the right amount of information. This means that a generic should be modified only if the property predicated is typical of the subclass denoted by the modified concept but not of the unmodified concept; otherwise, the modifier should have been omitted. The violation of this principle explains the prevalence of pragmatic effects observed in experimental studies on generics. When presented with a modified generic, listeners expect it to provide additional or different information from the unmodified version. If the predicated property is already expected for the unmodified version, they could interpret the modifier as pragmatic "noise" or as suggesting a contrastive reading (e.g., "Young lawyers are conservative" as opposed to "old lawyers"), leading them to draw pragmatic inferences.

Now, the perceived likelihood of a generic "*M*s are *X*" depends on the position of *X* in the expectation ordering *Exp*(*M*). In a modified generic "*NM*s are *X*," the modifier (*N*) induces changes in the original order of expectations because *NM* is a subclass of *M* with a potentially different property distribution. Consequently, its adequacy will depend on the (potentially different) position of *X* in the expectation order *Exp*(*NM*).

Following the example of Shafir et al. (1990, 237), people judge as more probable the statement "Apples are red" than "Sour apples are red." This is because the property *red* appears better positioned in the ordering *Exp*(APPLE) than in the modified ordering *Exp*(SOUR APPLE) because *red* is correlated with *sweet*. In other words, the prototype of the subclass SOUR APPLE is closer to the area of the conceptual space covering the property non-sweet than the prototype of the category APPLE. This generalizes to all atypical modifiers, especially when they are negatively correlated with the property in the generic.

Our model predicts that pragmatic factors, such as Gricean maxims, significantly influence the interpretation of modified generics, with the impact varying according to the typicality of the modifier. Some properties are so prevalent in a category that they go beyond mere typicality as they are strictly more expected than any other typical properties (appearing earlier in the order of expectations). Consequently, generics with highly typical modifiers are not interpreted as referring to meaningful subclasses of the head noun, resulting in pragmatically deviant expressions (Connolly et al. 2007, 10). Consider, for example, the modified generics "Adult penguins live in cold areas" and "Flightless penguins live in cold areas." While both *adult* and *flightless* are prototypical properties of PENGUIN, *flightless* is so prevalent that the combination *flightless penguin* does not refer to a meaningful subclass and thus seems odd. In contrast, ADULT PENGUIN does create a meaningful subclass, as not all penguins are adults (cf. Hampton et al. 2011, 103).[6]

5.6.3 Conjunction Fallacies

The above phenomena can be understood as *property verification tasks* (Solomon and Barsalou 2004) that exclusively involve knowledge-what. In other words, the category is fixed and we just check whether a property is more or less expected. There is, however, another group of reasoning "fallacies" that involve trying to predict category membership or correlated properties from given information about an individual. The most well known is the so-called conjunction fallacy in which people judge a conjunction of two events as more probable than one of its constituent events alone, violating the laws of probability (in particular, that for any two events A and B, $P(A \cap B) \leq P(A)$, and $P(A \cap B) \leq P(B)$). The classic demonstration of this fallacy is the "Linda problem" (Tversky and Kahneman 1983), where subjects were given the following information: Linda is thirty-one years old, single, outspoken, and very bright. She has a degree in philosophy, and while studying, she participated in anti-nuclear demonstrations and was involved with issues of discrimination and social justice. Afterward, subjects were asked which of the following statements is more probable:

1. Linda is a bank teller.
2. Linda is a feminist bank teller.

A large majority of the subjects answered that sentence (2) is more likely, although by the laws of probability, (1) is at least as probable as (2).

Several accounts have been proposed to explain the conjunction fallacy. Tversky and Kahneman (1983) originally attributed it to using the representativeness. However, multiple probabilistic solutions to the fallacy have been put forward after that. These include averaging models, where people are thought to average the probabilities of the constituent events instead of multiplying them (Fantino et al. 1997), and more complex rules, such as multiplicative models with configural weighting (Birnbaum et al. 1990).

Alternative explanations have also been proposed, such as the idea that the fallacy arises from random variation in probability judgments (Costello 2009) or considerations of source reliability in a Bayesian framework (Bovens and Hartmann 2003). In recent years, more psychologically plausible models have been advanced. Tentori et al. (2013) proposed a confirmation-theoretic approach in which the key determinant of the conjunction fallacy is not the perceived probability of the added conjunct but rather the degree to which the available evidence inductively confirms or supports it.

We propose a new way of analyzing the conjunction fallacy that relates to both the confirmatory approach and to Tversky and Kahneman's representativeness heuristic (see also Shafir et al. 1990). The fundamental idea behind our approach is that people perceive the task as a problem of partial information and pattern completion. The information given to subjects at the beginning of the task induces a partial representation of Linda. According to our framework, this corresponds to the construction of a partial vector in a multidimensional person space: Linda=<*woman*, . . . , *young*, . . . , *outspoken*, . . . , *studied_philosophy*, . . . , *socially_aware*, . . . , *bright*, . . .>. The ellipses in the vector stand for the domains of the person-space we do not have information about (we cannot assign a "value" to them). Subjects then form expectations about these uncertain values on the basis of some possible *pattern* suggested by the previous information and their own background knowledge. When presented with the alternatives of new information about Linda, people evaluate it in terms of how well they align with and complete the pattern established by their expectations and the given partial information.[7]

In the Linda scenario, the conjunction of properties "feminist bank teller" is preferred over "bank-teller" because it provides a simpler and more coherent completion of the pattern suggested by the description of Linda (see chapter 9 for an analysis of the notion of coherence). It aligns better with the values expected for the unknown dimensions of Linda's person vector, based on the given information and background knowledge about correlations between personal characteristics. Counterintuitively, this conjunction may be simpler in two ways: (1) It reinforces the established pattern, requiring fewer additional assumptions to reconcile with the given information. (2) In terms of information compression, it might allow for a shorter overall description of Linda by resolving implied uncertainties about her character. This suggests that people are intuitively seeking the most efficient representation of Linda, rather than strictly following probability rules. As explained by Chater and Vitányi (2003, 21), "The more structure people can find in a stimulus, the easier they find it to process and remember and the less random it appears."[8]

The first to analyze the conjunction fallacy in terms of defaults or expectations seems to have been Veltman (1998). He defined a collection X of properties as more representative of the category A than of category B if (1) for some properties P in X, P is a "normal" property of A but only an exceptional property of B, and (2) no property P in X is a "normal" property of B but only an exceptional property of A. Given this definition, the properties X ascribed to Linda are more representative of the category of feminist bank tellers than of the category of feminists. Veltman's solution is similar to ours in that he also bases it on the typicality of properties for a category. However, he exploits neither conceptual spaces nor any form of distance measure.

A related phenomenon, called the inverse conjunction fallacy, has been studied by Jönsson and Hampton (2006) in a series of experiments. Again, a generic (like "All candles are made of wax") is compared to a modified generic containing a conjunction of properties ("All expensive purple candles are made of wax"). The experiments show that, on average, participants judged the original generic acceptable more often than they judged the modified generic acceptable. Again, our model predicts this for the case where the prototype of the modified category EXPENSIVE PURPLE CANDLES is closer to the area of the conceptual space covering the property *not made of wax* than is the prototype of the unmodified category CANDLE.

5.7 Inferential Strength

When an inference is based on defeasible premises, a criterion for judging the strength of the inference is helpful. Deductive inferences, for instance, are considered maximally strong since they preserve truth. However, in everyday reasoning they can be analyzed as having different strengths according to the "quality" of the information in the premises. In other words, while the structure of a deductive inference guarantees logical validity, it doesn't preclude variations in informativeness—and ultimately "utility"—in everyday reasoning.[9]

Unlike deductive inferences, inductive inferences are not certain; the premises never fully guarantee the conclusion, meaning that inferring the conclusion always involves some degree of risk (see Hacking 2001). Since Carnap, inductive logicians have developed various methods to measure the degree to which premises support a conclusion in inductive arguments. These methods, primarily based on probability theory, treat the strength of the inferential relationship as an objective feature of arguments, captured by the concept of "evidential support."

From a psychological perspective, the perceived strength of an inductive argument depends on various cognitive and contextual factors. These might include the salience and diversity of premises, as well as the conclusion's believability. Category-based induction, for example, involves a complex interplay of factors related to conceptual similarity, typicality, and taxonomic organization (see chapter 7 and Osta-Vélez and Gärdenfors 2020).

In this chapter, we focused on a specific factor influencing the (subjective) strength of inductive inference: the degree of *entrenchment* of the premise information within the conceptual representation of the inference's main concepts. Recall that an expectation ordering is a type of epistemic entrenchment ordering. The notion of epistemic entrenchment, originating from belief revision theory, captures the idea that within a belief system, certain beliefs are more susceptible to revision or deletion than others (Gärdenfors 1988). In an expectation ordering, a belief's position indicates its comparative degree of defeasibility (Gärdenfors and Makinson 1994, 209). We propose using the typicality measure to determine a property's degree of defeasibility: The more typical a property, the less defeasible it is.

To give an illustrative example, suppose you are told that Maria has adopted a new pet and named it Rex. Given this information, you could infer that Rex flies if you use as an implicit premise from your set of expectations that Rex is a bird. However, this inference would be weak, since the concept BIRD would normally be less entrenched in *Exp*(PET) than concepts such as DOG or CAT. If, instead of BIRD, you use as an implicit premise that Maria's new pet is a dog or a cat, then you would infer nonmonotonically that Rex does not fly, and this inference would be stronger.

The above example shows that when we reason nonmonotonically, the strength of our inferences depends on the choice of information from the set of expectations we use as implicit premises. In more general terms, we can say that the strength of an inference is a negative function of the degree of defeasibility of the propositions that are used as implicit premises. As we mentioned before, our framework generates an explication of this criterion by considering that the information to be used is always the one present at the maximal point of the set of expectations—that is, the prototypical information of the concept in question or the one that results from an updating process as defined in section 5.4.

As explained earlier, even among the sufficiently well-expected information in the set of expectations, some propositions are less defeasible than others. As a result, they will produce stronger inferences. For a system of default logic, a comparative notion of inferential strength can be defined as a positive function of the strength of the default rules used in specific inferences. As we have shown, this strength measure comes from the position of the rule in the expectation ordering, which depends ultimately on the typicality measure. To give a simple example, an inference that uses the default $Bird(x) \rightarrow HaveFeathers(x)$ will be stronger than one that uses the default $Bird(x) \rightarrow Fly(x)$ since following the typicality measure, $HaveFeathers(x)$ will be more entrenched in the expectation order than $Fly(x)$ even if both pertain to the prototype of BIRD.

Finally, another important issue in default logic concerns the possibility of having defaults expressing conflicting information (Reiter and Criscuolo 1981; Touretzky 1984). The most famous example of this situation is the "penguin principle" (Lascarides and Asher 1993): If we are told that x is a bird and that it is also a penguin, then we will have at our disposal the defaults $Bird(x) \rightarrow Fly(x)$ and $Penguin(x) \rightarrow \neg Fly(x)$. If we do not have a clear criterion to choose which default has priority, we will end up addressing a contradiction. Gricean principles, however, entail that the more informative concept—that is, PENGUIN in this case—should have priority. Above, we mentioned the "Nixon diamond," where two conflicting defaults seemingly lead to the conclusion that Nixon is both a pacifist and not a pacifist. The typicality measure gives a way of resolving the conflict since once it is decided whether Nixon is more atypical as a quaker than as a republican, the resulting expectation ordering determines which of the default rules should yield.

5.8 The Meaning of "But"

We have expectations, but we are sometimes surprised. The choice of the word "but" instead of "and" in the previous sentence indicates that the information contained in the second half of the sentence *violates our expectations*. We shall argue that this is the core meaning of "but."

In introductory courses in logic, one often uses the formalism of propositional logic to analyze conjunctions in natural language. It is shown how connectors like "and," "or," "if . . . , then," "unless," and "even if" can be expressed in this formalism. However, "but" rarely receives a thorough analysis. At most, it is said to be synonymous with "and," a claim that any competent speaker of English would recognize as false. Among linguists, it is a commonplace that "but" expresses a violation of expectations.[10]

However, they leave it at that since they have no way of representing and analyzing expectations.

Using the tools of this chapter, the analysis of "but" we propose is the following:

(CBut) A sentence of the form "*p* but *q*" is acceptable in a context C if and only if *p* and *q* are both acceptable in C, and in C, it holds that $p \rightarrow \neg q$ is more expected than $p \rightarrow q$.

We don't propose any truth conditions for "but," simply because we do not think there are any. As will be clear soon, the use of "but" is highly context-sensitive, so we submit that the proper analysis of "but" should be in terms of the conditions under which a sentence is *accepted* in a given context.

The context also determines what the current expectations are. In our analysis, the use of "but" can be seen as applying the Gricean principle of maximal informativeness. If only *p* is uttered, then the hearer expects that the typical consequences of *p* also apply. To mark that the expectation $\neg q$ does not hold, the sentence "*p* but *q*" is expressed, following the principle of maximal informativeness.

Let us apply the analysis to some examples:

(5.1) He is rich and ugly.

(5.2) He is rich but ugly.

The difference between the content of (5.1) and (5.2) is that in (5.2), the speaker presupposes that rich men are normally not ugly, while such an expectation is not indicated in (5.1). The use of "but" *signals* an expectation.

The role of the context can sometimes be quite subtle. Compare the following two sentences (Lakoff 1971, 133):

(5.3) John hates ice cream but so do I.

(5.4) John hates ice cream, but I like it.

On a standard reading of (5.3), it carries the expectation that people normally like ice cream. The "but" indicates no contrast between me and John, but a contrast between me and people in general. However, in a natural context where (5.4) is uttered, there is no expectation concerning how normal people like ice cream. On the other hand, mentioning John's dislike for ice cream creates an expectation, albeit a weak one, that I too should dislike it, and it is this "inductively generated" expectation that is denied by the "but." The same argument applies to another of Lakoff's examples:

(5.5) John is tall, but Bill is short.

She calls examples like (5.4) and (5.5) instances of "semantic oppositions" and argues that it is one of the functions of "but" to express such oppositions (Lakoff 1971, 133). However, we find it more natural to view sentences (5.4) and (5.5) as a special kind of denial of expectations, which is the other meaning of "but" that Lakoff identifies and which is expressed more formally in our analysis above.

In this section, we have outlined an analysis of the meaning of "but" based on expectations. In addition to this application, we believe that a large part of the discussion within linguistics and philosophy concerning *presuppositions* of sentences can be given a more unified treatment in terms of the speaker's expectations. However, such an analysis will not be attempted here.

5.9 Conclusion

Despite the significant progress made in the field of nonmonotonic reasoning during the past decades, the available logical models still suffer from various epistemological issues. Several formal systems make strong assumptions about the role of background knowledge and entrenchment relations of beliefs without providing any cognitive or epistemological arguments for them.

In this chapter, we have argued that some of these problems are due to the intrinsic limitations of propositional-based models for capturing the internal structure of conceptual knowledge. We have shown that a framework that interprets nonmonotonic inference as a form of expectation-based reasoning, with expectations grounded in conceptual spaces, fruitfully extends the modeling tools of these logical systems while enriching their theoretical foundations. Furthermore, we demonstrate the potential of this approach by showing how it can account for a range of inferential phenomena often described in the psychological literature as logical or probabilistic fallacies.

There exist attempts to computationally implement reasoning with default rules (Lieto and Pozzato 2018; Brewka et al. 2007; Delgrande and Schaub 2000). However, in these systems, the default rules must be provided by the user, and they do not generalize well. We submit that a conceptual spaces approach to modeling reasoning with default and nonmonotonic reasoning is a better method when attempting to build artificial systems with these capacities. Once the domains with their concept regions and distance functions have been implemented (we are not saying that this is an easy task—see section 2.9), then the typicality measure provides a direct way to generate expectation orderings and default rules, and thereby a method to calculate nonmonotonic inferences. Implementations, however, remain to be constructed.

The ideas defended here are directly related to a kind of reasoning known in psychology as category-based induction. This mechanism consists of judging the strength of arguments that project some property from one or more categories to another category, exploiting conceptual similarities (see Feeney 2018). This will be the topic of chapter 7. We believe that explicating nonmonotonic reasoning and category-based induction with the same modeling framework is valuable for the various disciplines studying these phenomena and contributes to the systematization of the formal frameworks used in theories of reasoning.

Finally, one limitation of the research presented here is that the expectations we model concern only object concepts. However, expectations are pervasive in cognition, and a full model of their role in reasoning should also account for more complex lexical items like verbs or relational and functional concepts. In Gärdenfors (2014, 2024a), the conceptual space framework has been extended for modeling expectations based on the structure of events, in particular causal inferences. Future work on expectation-based reasoning will hopefully result in a more general model.

6 Generics

Throughout this book, we have identified some limitations of logical approaches to reasoning, arguing that analyses confined to sentence structure are insufficient to explain rational inference. In this chapter, we extend these arguments. Specifically, certain expressive formats in natural language have logical forms that are either ambiguous or unrecoverable, yet they still play a crucial role in reasoning and communication. This poses a further obstacle for those who apply strictly logical frameworks to analyze human reasoning (see Oaksford and Chater 2002).

We focus on generic sentences, a construction in natural language used to convey generalizations without using quantifiers. Expressions of the form "*F*s are *G*"—for example, "Birds fly," "French like wine," or "Tigers are ferocious"—are typical cases of generics. From an epistemological perspective, generics are fundamental because they serve as a vehicle for communicating world knowledge and play crucial roles in learning and everyday communication (Krifka et al. 1995; Gelman 2003, 2010; Leslie and Lerner 2016). Interestingly, children learn to use generics earlier than universal quantification, a fact that speaks to their importance in cognitive development.

Generics are also central to everyday reasoning. Although they resemble universal statements conferring extensional information, they differ in an important way: Generics express knowledge-what, while universal sentences typically express knowledge-that. As discussed in the following section, this has important implications for how we understand and use them in reasoning.

In this chapter, we show how generics can be explained in the framework of our theory while providing an analysis of their role in everyday reasoning.

6.1 Why the Meanings of Generic Sentences Have Been Elusive

On the surface, generics seem to be about the prevalence of properties in groups of individuals. However, when we analyze how they are used in everyday cognition, we see that the prevalence factor plays a minor role. Consider the following statements:

(6.1) Tigers are striped.

(6.2) Ducks lay eggs.

(6.3) Sharks kill people.

(6.4) Bees are sterile.

Sentence (6.1) is unproblematic since almost all tigers are striped; (6.2) is often considered to be true, although the property in it applies only to a minority of ducks (mature female ducks); (6.3) is also generally accepted, despite the fact that human deaths caused by sharks are quite rare and most shark species are harmless; finally, (6.4) is false, although the property "sterile" is much more prevalent for bees than are the properties in (6.2) and (6.3) for ducks and sharks, respectively. This shows that generics form a heterogeneous class that is difficult to express coherently by statistical or quantificational rules.

Within analytic philosophy, there have been several attempts to analyze generics by studying their truth conditions (Pelletier and Asher 1997; Cohen 2004; Sterken 2015b; Van Rooij and Schulz 2020). Philosophers often argue that generics are not real universals but implicitly mean something like "Typically, birds fly" or "Usually, penguins do not fly." This has led to the suggestion that generics have a "deep structure" that combines two predicates with a hidden operator—often called *Gen*—that refers to an adverb of quantification like "typically," "generally," or "usually."

If this were the case, generics would have a proper logical form, like $Gen(x)$ $(Bird(x) \rightarrow Fly(x))$, making them analyzable in terms of truth-functional structure (Chierchia 1995; Pelletier and Asher 1997). This approach, however, faces several technical problems and fails to capture the role of generics in everyday communication and reasoning (see Leslie [2008] and Leslie and Lerner [2016] for arguments).

In this chapter we defend an alternative view: Generics are related to the *intensions* of concepts rather than to their *extensions* (Cimpian et al. 2010; Hampton 2012; Leslie 2012). More specifically, generic sentences express expectations about *relations* between object concepts and properties and, therefore, a form of knowledge-what. We shall argue that they cannot be evaluated as isolated sentences but only in association with clusters of sentences encoding conceptual knowledge. Instead of looking for truth conditions, we focus on how generics are used, that is, their role in reasoning, learning, and communication.

We need a new methodology to build computational models of reasoning with generics, and our shift from truth-conditional approaches opens the way for this. Although this chapter doesn't present specific algorithms, it builds on the framework for reasoning with expectations introduced earlier, laying the groundwork for developing alternative computational methods.

6.2 Property Versus Diagnostic Generics

The classification of generics has been the subject of extensive debate in the literature. For example, Leslie et al. (2011) and Prasada et al. (2013) propose at least five different types.[1] We herein advance a new classification. The first pertinent distinction is between *normative* and *descriptive* generics. Descriptive generics, such as "Birds fly," state knowledge of factual patterns, whereas normative generics, such as "Boys don't cry," express social expectations. Our analysis focuses exclusively on descriptive generics and classifies them into two types: *property generics*, which refer to characteristic properties of objects, and *diagnostic generics*, which refer to properties that are salient in a category in light of the properties of categories in their contrast class.

A subset of generics termed *striking generics*, such as "Mosquitoes carry the West Nile virus" and "Sharks kill people," has garnered significant attention (Leslie 2008; Prasada et al. 2013; Sterken 2015a; Van Rooij and Schulz 2020). We will show that striking generics can be analyzed in terms of diagnosticity and affective valence, and we will discuss the extent to which they constitute an additional subtype of generic sentences.

In section 6.3, we employ the theory of conceptual spaces to explain how generics reflect structural features of conceptual knowledge, particularly its prototypical structure. Building on the analysis presented in chapter 5, we also explore the role of generics in expectation-based reasoning, drawing from the work of Gärdenfors and Makinson (1994) and Osta-Vélez and Gärdenfors (2021).

Section 6.4 focuses on property generics while also explaining a type of generics that involve hierarchies of concepts. In section 6.5, we turn our attention to diagnostic generics—statements such as "Lions have manes"—that appear to attribute typical properties but in fact attribute diagnostic properties that mark contrast with other classes. We argue that diagnostic generics are fundamentally different from property generics and demonstrate how our measure of diagnosticity can be used to evaluate this type of generic sentence. Additionally, we briefly compare diagnostic generics to those expressing striking properties, such as "Sharks kill people." Finally, in section 6.6, we discuss the pragmatics of generics, with a particular focus on their role in teaching.

6.3 Generics as Expectations

6.3.1 Background Knowledge and Expectations

Within analytic philosophy and linguistics, it is common to understand the meaning of a sentence in terms of truth conditions (see section 1.2). In the case of generics, however, a functional approach might be more suitable (Greenberg and Harman 2006). From this perspective, the meaning of sentences depends on their roles in reasoning and communication. As a result, sentences cannot be understood in isolation but must be interpreted in relation to background knowledge and pragmatic factors, such as the communicative intentions of agents.

Based on this functional approach, we propose our first thesis: The meaning of a generic sentence can only be evaluated in light of an agent's background knowledge. This thesis applies to two types of agents involved in the use of generics. The first type is an agent expressing a generic sentence as information for someone else. The second type is an agent evaluating a generic after hearing or reading it. In the main part of this chapter, we focus on the first type of agent. We will return to the second type in section 6.7.

As we explained in chapter 5, expectations are crucial for reasoning under uncertainty, functioning as implicit premises that allow us to draw inferences when complete information is unavailable. Building on this idea, our aim now is to demonstrate that generics can be viewed as expectations and that they play a significant role in communicating our conceptual commitments. Instead of discussing generics as statements that can be either true or false, we will discuss them as having different "degrees of strength." This will be done in terms of the *typicality* and *diagnosticity* of properties of concepts.

6.3.2 The Internal Structure of Generics

Generics, in their paradigmatic form as bare plurals, relate a property *P* to some object concept *C*. Consequently, the evaluation of a generic statement depends on our knowledge of *P* and its interaction with our understanding of concept *C*. Surprisingly, few theories of generics focus on this interaction, and those that do (Declerck 1991; Heyer 1990) lack a substantive theory of background knowledge or concept structure. To address this, we employ the distinction between knowledge-that and knowledge-what discussed in previous chapters, providing a framework for analyzing the relations between object concepts and properties that largely determine the content of a generic (see Prasada et al. 2013).

The central idea is that property generics (as sentences expressing knowledge-what) are evaluated by considering different compatibility relations between the semantic domain of the property and the object concept in the sentence. In brief, generics concern more or less prototypical or diagnostic properties of concepts. As we will see, these ideas can be made more precise by using conceptual spaces as a framework for modeling the structure of concepts.

6.3.3 Generics, Reasoning, and Expectations

Generics derive their meaning from a complex web of background knowledge, and their interaction with this body of information significantly influences how we use them in reasoning. Consider, for example, the generic "Birds fly." On its own, it is totally acceptable. However, introducing a taxonomic generic such as "Penguins are birds" can create a logical conflict. If we apply the generic "Birds fly" to penguins, it would suggest that "penguins fly," which contradicts what we know about penguins—that they do not fly.

In this scenario, the generic "Birds fly" is weaker than the more specific generic "Penguins do not fly." Flying is not a prototypical property for penguins, while non-flying is. Therefore, when we reason about penguins, we tend to ignore or discard the generic "Birds fly" to avoid conflict and stick with the fact that penguins do not fly. Other generics that fit better with what we know, such as "Penguins eat fish," can still apply. This process of resolving conflicting information follows principles governing expectations that were presented in the previous chapter (Gärdenfors and Makinson 1994; Osta-Vélez and Gärdenfors 2021).

A general principle when reasoning with dynamic information is the following: When someone begins with the sentence "*x* is *C*" and later receives the new information "*x* is *D*," compatible with *C*, her uncertainty regarding the expected properties of *x* will be reduced and she will be able to make more precise inferences with her background knowledge. A paradigmatic case of this is when *D* is a subordinate concept of *C* (like PENGUIN and BIRD). In the conceptual spaces framework, reducing uncertainty about an object *x* implies reducing the *volume* of the subregion of the space in which *x* is located.[2] Such a reduction, however, may make new properties prototypical (*not flying* is prototypical for PENGUIN, but not for BIRD). This analysis shows that reasoning with generics has nonmonotonic properties since adding more information to the given background knowledge may result in some generics no longer being accepted.

6.4 Representing Property Generics in Conceptual Spaces

6.4.1 Property Generics and Typicality

Because generics express expectations and expectations can be ordered by their degree of entrenchment in a concept (as shown in the previous chapter), generics themselves exhibit a graded nature and can also be ordered. In particular, we claim that the strength of a property generic will be a positive function of the *typicality* of the property within a conceptual space.

The general rule is that the more typical a property is for a particular object concept, the more expected the generic. This rule explains why the proportion of instances is not decisive for how useful a generic is in arguments. For example, "Books are paperbacks" describes a highly prevalent feature of books but does not concern a characteristic property.

From our perspective, a way to determine such an expectation ordering is to consider the relative typicality of each property in its conceptual space. For instance, *round* and *smooth* are more typical properties of APPLE within a fruit space than *red* since instances of non-red apples are more typical than non-round or rough apples (see figure 5.1). This means that the former two properties will have priority over the latter in an expectation order.

The most important consequence of this analysis of how generics are used in reasoning is that it does not require any additional linguistic or logical operators. No *Gen* or other default operators are needed. In support of this, it can be noted that there is no known language that has a dedicated, articulated generic operator. Instead, we assume that the acceptability of generics can be determined by the strength of their expectations within the appropriate conceptual space.

Our analysis thereby explains why it is easier for children to learn to use generics than quantified sentences. There is evidence that generics are the default form of generalization children use when reasoning about kinds (Gelman 2010; Gelman et al. 2015). According to our construal, these generalizations depend exclusively on the organization of conceptual knowledge, which is already present in the early stages of development (Mattos and Hinzen 2015). In contrast, mastering quantifiers does not seem to rely on our knowledge of concepts but on forms of extensional reasoning that are less intuitive for children (Gelman 2010). Thus, understanding generics as *knowledge-what* seems more natural for explaining their prevalence in infant cognition than an extensional view (knowledge-that) that assumes the existence of implicit quantifiers in these sentences. We expand on this in section 6.7.

Let us now focus on the graded nature of generics, an aspect that has surprisingly not been studied in the experimental literature. Consider the following two generics:

(6.5) Elephants have trunks.

(6.6) Elephants are gray.

The property "having a trunk" is more characteristic of elephants than being gray. One can quite easily accommodate the occurrence of a white or a black elephant, and perhaps also a pink one. However, an elephant without a trunk is an incomplete elephant and is much less expected than a non-grey elephant. In terms of prototypes, a non-grey elephant is more similar to the prototypical elephant than an elephant without a trunk.

The general rule is that the more typical a property is for a particular object concept, the more expected the generic. This rule explains why the proportion of instances is not decisive for how useful a generic is in arguments.

Our key idea is to interpret a property generic "*C*s are *P*" as "*C*s that are similar to the prototype have the property *P*." The *strength* of such a generic can be determined by measuring the distance to the closest point where the property *P* is *not* satisfied. We use the typicality measure in the conceptual space to obtain this kind of information. For example, the distance from an elephant that does not have a trunk to a prototypical elephant is larger than the distance from an elephant that is not gray to a prototypical elephant. According to our criterion, (6.5) is therefore a stronger generic than (6.6).

When information about distances in a conceptual space is available, the criterion for measuring the strength of generics we propose can be used to make testable predictions concerning how subjects will judge generics. Some general principles can also be derived. For example, a generic "Cs have the property P&Q" will have the same strength as the weakest of "Cs have the property P" and "Cs have the property Q." This is so because the closest point where P&Q is not satisfied is the same as the closest where either P or Q is not satisfied.

6.4.2 Two Tests for Generic Statements

It is well known that property generics and factual statements behave differently in language. One test is proposed by Lawler (1973):

(6.7) Blue whales eat plankton.

(6.8) A blue whale eats plankton.

(6.9) Blue whales can be seen around the Cape of Good Hope.

(6.10) *A blue whale can be seen around the Cape of Good Hope.

(6.7) describes a characteristic property of BLUE WHALE. It can be exchanged for the indefinite singular version in (6.8). The generic expresses a relation between the concept BLUE WHALE and the property of *feeding on plankton*. In contrast, (6.9) is a factual statement. A test for this is that it cannot be exchanged for the indefinite singular version in (6.10) (Carlson 2009; Krifka 2012). As we will show in section 6.6, this test only applies to property generics, not to diagnostic generics.

A second test for the difference between generics and factual sentences is that generics are not "upward entailing." This means that in (6.7), BLUE WHALES cannot be replaced by the superordinate category MAMMAL since "Mammals eat plankton" is not acceptable. On the other hand, factual sentences are upward entailing: (6.9) entails "Mammals can be seen around the Cape of Good Hope."

The fact that sentences (6.7) and (6.8) express the same content despite their different logical forms further indicates that generics form a special class of sentences. The upshot is that although a generic is a sentence, it expresses a different kind of knowledge than a factual sentence—that is, knowledge-what rather than knowledge-that.

6.5 Diagnostic Generics

The typicality measure provides a solid platform for understanding the acceptability conditions of a property generic. However, generics can also be evaluated from an informational perspective. To do so, we must shift our attention from the *internal* to the *external* structure of object concepts—that is, to superordinates and *contrast classes*.

6.5.1 Knowledge About Superordinates

We mentioned in section 2.5.2 that concepts can be understood as organized in a "horizontal" and a "vertical" dimension (Rosch 1988). The subordinate-superordinate relation is encoded in the "vertical" dimension. We will refer to the contrast class of category *X* as the category *CC*(*X*) containing the categories that are different from *X* but share its immediate superordinate category with it. For example, *CC*(LION) is FELINE.

A significant part of our expectations about an object concept is *inherited* from knowledge of its superordinate. For example, one may have minimal knowledge of what a lynx is, but merely knowing that it is a feline will generate expectations (of varying strength) derived from the defining properties and characteristics of that superordinate concept. We submit that inherited expectations play an important role when evaluating generics with minority characteristic properties.

Consider the following generics:

(6.11) Birds lay eggs.

(6.12) Bees are sterile.

(6.13) Lions have manes.

From an extensional point of view, it seems puzzling that (6.11) is true while the property only applies to a minority of the individuals of the class (adult females). On the other hand, (6.12) is false, but the specified property applies to a large majority of the individuals in the category. The confusion disappears when we consider these generics in interaction with inherited expectations.

Since *egg-laying* is prototypical for BIRD, the typicality criterion offers a straightforward explanation of why (6.11) is usually accepted as true. However, the informational relevance of (6.11) depends on its interaction with the set of inherited expectations {"Birds are animals," "Animals reproduce"}, which consequently entail "Birds reproduce." By specifying a mode of reproduction for BIRD, (6.11) adds relevant information to x and thus reduces the degree of uncertainty in (the inferentially available generic) "Birds reproduce."

Similarly, we can explain why (6.12) is usually taken to be false by appealing to the fact that the property in it is not prototypical of BEE. However, this explanation can be complemented from an informational point of view if we note that (6.12) can be interpreted as implying that "bees cannot reproduce," a generic that conflicts with the set of inherited expectations {"Bees are animals," "Animals reproduce"}, which have as a consequence that bees reproduce.[3]

6.5.2 Applying the Measure of Diagnosticity

Many generics express properties that are diagnostic for a concept. A property is diagnostic when it maximizes the dissimilarity between an object concept and the concepts in its contrast class, speeding up categorization (see section 2.7.3). For example, in (6.13), having a mane is diagnostic for lions since having manes distinguishes them most (at least their adult males) from other felines. In other words, *having a mane* is diagnostic for LION in *CC*(LION) = {TIGER, CHEETAH, JAGUAR, LEOPARD, CAT, and so on}. In other words, (6.13) can be read as "In contrast to other felines, lions have manes."

In our model, we address the influence of the contrast class by considering the prototype of the immediate superordinate concept. In the case of (6.13), explaining why it is an informationally relevant generic involves considering its relationship to the contrast class *CC*(LION). More specifically, the informational value of (6.13) is related to the fact that the property "has a mane" is not present in any other of the concepts in *CC*(LION) and, therefore, is highly diagnostic for LION. Recall that we define a measure of diagnosticity as follows: Given an object concept *M,* its immediate superordinate concept *N,* and a property, the diagnosticity of R_i is equal to $\frac{T_M(R_i)}{T_N(R_i)}$, where $T_M(R_i)$ expresses the typicality of property in relation to the superordinate.

Diagnosticity is important since diagnostic properties carry the most information about an object category compared to similar concepts. So, for example, when children learn about a new object category, being told about such properties will speed up their learning (Gelman 2003).

We predict that generics with diagnostic properties are easier to endorse than generics with non-diagnostic ones because more information is added to an agent's conceptual systems. For instance, (6.13) should be seen as more interesting than "Lions have whiskers" because *having manes* contains information that is specific to the object category and helps to differentiate it from other concepts while *having whiskers* is common to all members in *CC*(LION) and can be inferred from characteristic knowledge about the superordinate FELINE.

Diagnosticity also comes in degrees. Some properties can have diagnostic value for an object category and still be non-prototypical or rare. Consider the generic:

(6.14) French people eat horse meat.

Horse meat is atypical in the diet of most French people (although it is common in a few regions of the country); Why, then, is (6.14) considered an acceptable generic? The answer is that the diagnostic value of the property in (6.14) in relation to *CC*(FRENCH PEOPLE) compensates for its low degree of typicality. In other words, the property *eat horse meat* will be more expected for FRENCH PEOPLE than for any other category in *CC*(FRENCH PEOPLE).

The diagnosticity measure assures that for any two similarly distinctive properties of an object concept (with respect to its contrast class), the one with the highest typicality degree will be the one with more diagnostic value. This allows us to capture the intuition that generics with prototypical properties with diagnostic value are stronger than generics with atypical properties that also have diagnostic value: For instance, "French people speak French" is stronger than "French people eat horse meat."

6.5.3 Generics with Striking Properties

In the literature (Leslie 2008; Prasada et al. 2013; Sterken 2015a; Van Rooij and Schulz 2020), so-called striking generics have been brought forward as a special type of generics. Two examples are the following:

(6.15) Ticks carry Lyme disease.

(6.16) Sharks kill people.

Such generics are problematic for any quantificational approach since only a small minority of ticks carry Lyme disease and shark attacks are very rare.

In our analysis, striking generics are special cases of diagnostic generics. (6.15) expresses that even though very few ticks carry Lyme disease, this is more frequent in ticks than in any other category in CC(TICKS) (the superordinate may perhaps be BUGS—ticks are not insects but arachnids). Ticks may, in fact, be the only animals that carry the disease. Similarly, even though sharks' attacks are rare, encounters with sharks are more dangerous than encounters with most other animals.

The linguistic form of a striking generic is generally a combination of a noun phrase and a verb phrase that is not based on "is" or "has." Indeed, all the striking generics in the tests of Prasada et al. (2013) contain such verbs (see table A on p. 420).

Although they are ultimately diagnostic generics, striking generics could form a special subclass of generics if we consider their affective dimension. Since striking generics refer to properties or dispositions that are "dangerous and to be avoided" (Prasada et al. 2013, 409), they must bear negative valence; that is, they elicit negative emotions and consequently draw more attention. This phenomenon aligns with the "automatic vigilance hypothesis," which posits that negative stimuli command more attention than neutral or positive stimuli (Pratto and John 1991; Öhman et al. 2001). This automatic attentional bias toward negative information makes such statements more memorable and, consequently, more likely to be acceptable despite the rareness of the events.

6.5.4 Comparing Property and Diagnostic Generics

Our analysis of property generics and diagnostic generics suggests that there are two factors at work in the evaluation of a generic.[4] The first has to do with the degree of typicality of the property in the generic, and the second with its informational contribution—that is, how diagnostic the property expressed in the generic is. To illustrate the difference between them, compare (6.13) with the generic "Lions have bones." Although very strong, the latter generic may seem too obvious since the information in it is inferentially available from the fact that lions are mammals. That is, minimal knowledge about the object concept LION allows you to infer that generic. Thus, it is acceptable but not pragmatically relevant. On the other hand, (6.13) contains information that cannot be inferred from the superordinate of LION and that contributes to the specification of the object concept in relation to its contrast class.

This informational factor in the evaluation of generics can be explained in terms of Wilson and Sperber's notion of "relevance":

> When is an input relevant? Intuitively, an input (a sight, a sound, an utterance, a memory) is relevant to an individual when it connects with background information he has available to yield conclusions that matter to him [. . .]. In relevance-theoretic terms, an input is relevant to an individual when its processing in a context of available assumptions yields a positive cognitive effect. (Wilson and Sperber 2004, 608)

Generics that cohere with our expectations and provide information that reduces their uncertainty have a positive cognitive effect and are, therefore, relevant in communication. Generics that provide concept-specific information that helps to separate an object concept from the concepts in the contrast class are also relevant in a similar sense. On the other hand, generics that potentially contradict our expectations or induce false conclusions are irrelevant and, therefore, often rejected.

Like property generics, diagnostic generics are not upward entailing: "Lions have manes" does not entail "Mammals have manes," and "Turtles have shells" does not entail "Reptiles have shells." Hence, this property is a test for determining whether a sentence is a generic and also for causal generics.

In contrast to property generics, however, diagnostic generics do not pass Lawler's (1973) test: The generic "Frenchmen eat horse meat" does not express the same as "A Frenchman eats horse meat," and similarly, "Lions have manes" is different from "A lion has a mane." This observation supports that the two types of generics are indeed different. The differences between property generics, diagnostic generics, and factual sentences are summarized in table 6.1.

Table 6.1
Comparing property generics and diagnostic generics

	Property generics	Diagnostic generics	Factual sentences
Passes Lawler's test	Yes	No	No
Upward entailing	No	No	Yes

6.6 The Pragmatics of Generics

So far, we have only considered generics from the perspective of an agent that evaluates its acceptability, given the background knowledge and expectations of the agent. In brief, our answer is that the acceptability of a generic is determined by prototypical properties and distances in a conceptual space. In this section, we turn to the second type of agent: somebody who hears or reads a generic. The question is how it will affect the agent's cognitive state—that is, how a new generic will change the agent's state of knowledge. We thereby turn to further pragmatic aspects of generics. In particular, we focus on the question: What is the *use* of generics in social interactions? (Gelman 2021).

We speculate that children at an early age learn to reason with the expectations that are generated by the generics they hear, but we know of no empirical investigations related to this position. Novel nouns provided by caregivers are interpreted as inference-rich, thus denoting coherent concepts (see chapter 9). This assumption helps children figure out which object concepts are kind concepts (Gelman 2003, 184, Cimpian and Scott 2012). Children expect the first count nouns they learn to refer to basic-level concepts.

The nouns then appear in generics (both property generics and diagnostic generics) together with an adjective that denotes a typical or diagnostic property of the noun. Gelman (2003, 174) writes that generics have two main functions for children: Firstly, generics teach children particular concept-wide generalizations. Secondly, they indicate that members of an object concept are alike in important ways. Gelman and Roberts (2017) write that "labels and generic language ensure speed, fidelity, and ease of transmitting category information by means of an overt and stable representational format" (7904).

Several studies argue that young children understand generics and can distinguish them from non-generics based on several types of cues. In particular, Gelman (2010) shows that by about the age of two and a half, children start producing generics, and at the age of four, generics constitute 3 percent of the sentences produced. This may not seem a high rate, but it should be considered in the context of all other sorts of utterances that children can make. They rely on generics when they learn about categories and use them to draw inferences. Children also interpret generics as different from quantificational sentences containing "all," "some," or "most."

Generics have a central role in *teaching,* particularly in what is called "natural pedagogy"—that is, teaching by parents and others in everyday circumstances (Csibra and Gergely 2009). We tell our children, already when they are small, things like "Cats say meow, dogs say woof, and cows say moo."[5] Later in school, they learn generics like "Tigers have stripes," "Copper conducts electricity," and "Democracies have freedom of speech." Such property generics are a way of presenting characteristic properties of various concepts (Leslie 2008). Learning about object concepts is primarily done via their characteristic properties.[6] And when it comes to diagnostic generics containing striking properties such as "Dogs bite people," they function as guidelines for caution in actions (Sterken 2015a).

Mattos and Hinzen (2015) argue that natural pedagogy is one of the main functions of language. They write that humans have a "specific capacity to acquire, through communication, different kinds of information—respectively, *knowledge about kinds* and knowledge about particular events, actions, and state of affairs, which we will call here simply *knowledge about facts* (Mattos and Hinzen 2015, 7). This distinction is the same as the one we make between knowledge-what and knowledge-that. They also argue that children learn knowledge about kinds earlier than they learn knowledge about facts.

Gelman and Roberts (2017, 790) analyze the role of *labels* in cultural inheritance. They argue that "category labels work in an almost paradoxical way to ensure stability in the transmission process, but simultaneously to permit and even foster conceptual change." Because young children act as if their own knowledge state is the same as that of others, this will improve learning. Sabbagh and Henderson (2007) also argue that children's understanding that the meanings of words are conventional makes their word learning more efficient. They write that "these limitations in children's abilities to reason about others' epistemic mental state may actually promote the development of an appreciation of conventionality" (Sabbagh and Henderson 2007, 33). Labels for objects also license inferences that depend on underlying causal features of objects. Such connections are taught via generics.

What has been called the *principle of conventionality* (Clark 1993) states that words are efficient tools for communication when the form–meaning associations are known, shared, and expected within a language community. As young as one-and-a-half-year-old children exploit this principle in their learning of language (Graham et al. 2006). A reasonable explanation is that they believe that labels reflect what objects are called rather than what people call objects.

Another way of supporting the idea that the pragmatics of generics mainly concern teaching is to consider the *evolution* of teaching. Gärdenfors and Högberg (2017, 2024) distinguish between six levels of intentional teaching: (1) intentional approval/disapproval, (2) drawing attention, (3) demonstrating, (4) communicating concepts, (5) explaining concept relations, and (6) narrating. Of particular interest in this context are levels (4) and (5). As regards communicating concepts, Gärdenfors and Högberg (2024) write, "In human societies, the main methods to teach a concept is to use a word, sound or gestural sign standing for the concept together with pointing or some other technique for drawing the attention to what characterizes the category" (cf. Mattos and Hinzen 2015).

And level (5)—explaining concept relations—is introduced as follows: "The teacher's intention in explaining is typically that the learner understands the causal relationship between two or more concepts. For example, if this arrow is dipped in poison, it will kill the animal that it hits" (Gärdenfors and Högberg 2024). These descriptions were written without any comments on generics, but it is clear that levels (4) and (5) relate to generics. Although speculative, this connection could lead to an explanation of why all languages have generics and why they exhibit the two types that we have distinguished.

6.7 Conclusion

In this chapter, we have put forward the following theses:

(1) Generics should be seen as defeasible expectations that are added to the background information when reasoning.

(2) Generics express knowledge-what, not knowledge-that.

(3) There are two main types of generics: property generics and diagnostic generics.

(4) Property generics are evaluated in terms of typicality.

(5) Diagnostic generics are evaluated in terms of diagnosticity.

The measures of typicality and diagnosticity determine the strength of different generics. The upshot is that the measures of typicality and diagnosticity are also useful tools when judging generics. Furthermore, once information about distances in a conceptual space is available, the measures can be used to generate testable predictions concerning how subjects will judge the strength of different generics. And if distances are available, the typicality measure could be used to computationally determine the strength of different generics, thereby placing them in an expectation ordering. Following the method in chapter 5, such an ordering could then be used to include generics when reasoning in artificial systems.

7 Category-Based Induction

7.1 What Category-Based Induction Is

Induction and deduction deal with semantic information in different ways (Johnson-Laird 1983). In deduction, the conclusion follows with certainty, since the information it conveys is already contained in the premises. In induction, by contrast, the conclusion adds information that is not contained in the premises and thus introduces uncertainty. The textbook example of an inductive inference illustrates this by showing how we generalize a pattern from a limited set of observations—for instance, inferring that all ravens are black based on seeing only some ravens. But much of the induction we engage in daily does not depend on just counting cases, but builds on expectations and background knowledge to go beyond the data at hand. In this chapter, we argue that a complete understanding of induction requires to explain how prior knowledge and expectations are combined in reasoning to yield informative and reasonable inferences under uncertainty. In line with the proposal developed in the previous chapters, we suggest that this can be done if inference is conceived as a process that combines information codified at the symbolic (linguistic) level with information encoded as knowledge-what at the conceptual level.

In his pioneering article, "Inductive judgments about natural categories," Rips (1975) analyzed a type of inductive reasoning that exemplifies the idea above; in it, people use knowledge of concepts and their relationships to estimate the likelihood of projecting properties from one category to another. For example, they might evaluate an argument like "Dogs have sesamoid bones; thus, wolves have sesamoid bones" on the basis of the conceptual similarity between DOG and WOLF—rather than the logical form of the argument or some other propositionally codified property. This mechanism, known as *category-based induction* (CBI), plays a fundamental role in our cognitive lives by helping us manage uncertainty and by making cognition more efficient (Feeney 2018, 167).

Understanding how CBI operates, particularly which features of our conceptual systems it relies on, offers valuable insights into the broader role of concepts in reasoning.[1] In this chapter, we explore the general characteristics of CBI and propose a model that builds on distances in conceptual spaces. We demonstrate that this model accounts for most of the empirical findings related to CBI.

The chapter is organized as follows. Section 7.2 presents a basic taxonomy of category-based inductions and reviews the central experimental results associated with them. Section 7.3 presents our model and explains how the theory of conceptual spaces provides a natural way of modeling the central properties of CBI based on the theory's capacity for representing similarity and typicality relations among categories. Section 7.4 compares our model to some previous explanations, and section 7.5 presents some experimental tests of the model. Finally, section 7.6 briefly considers some methodological aspects of the approach we defend.

This chapter is technically more complicated than the others. However, the similarity and typicality measures prove their mettle here, since we can use them to explain a wealth of psychological evidence concerning CBI. So bear with us!

7.2 Types of Category-Based Induction

7.2.1 The General Structure of Category-Based Induction

Rips's (1975) paper sought to understand people's strategies when reasoning under uncertainty about animal concepts and their properties. He showed that agents exploit structural properties of concepts to estimate the likelihood that a certain property of one concept will extend to another.

In particular, Rips identified two guiding principles in this type of reasoning: similarity between concepts and prototypical structure within a concept.

Most studies on category-based induction (CBI) build on Rips's analysis (Osherson et al. 1990; Sloman 1993; Carey 1985; Heit 1997), positing that CBI relies on information at the conceptual level rather than the propositional level. Methodologically, these studies draw conclusions about cognitive agents' inferential tendencies by examining how individuals assess the strength of various types of categorical induction. Numerous CBI phenomena have been identified in recent decades (see Hayes et al. [2010] and Feeney [2018] for reviews). We now present a basic taxonomy of the phenomena that characterize CBI.

Category-based inductions are structured as arguments with one or more premises of the form "*X* are *S*" (e.g., "Dogs have sesamoid bones" or "Bears love onions") and a conclusion of the same type. We abbreviate an inference of the form "*X* have property *S*; thus, *Y* have property *S*" as $X \rightarrow Y$. This abbreviation is appropriate because most experiments investigate inferences with "blank" properties—where participants possess little or no knowledge about the property *S*—ensuring that this lack of knowledge does not affect the strength of the argument.

CBI can be classified in two major ways: by the number of premises and whether the conclusion belongs to the same conceptual level as the premises or to a superordinate concept. When the premise(s) and conclusion object concepts are at the same conceptual level, the argument is called *specific*; when the argument involves a "jump" to a superordinate level, it is called *general*. For instance, arguments with the form ROBIN → CROW or TABLE → CHAIR are specific, while arguments that generalize to a superordinate concept, like ROBIN → BIRD, ROBIN → ANIMAL, or TABLE → FURNITURE, are general. Both specific and general arguments can be composed of one or multiple premises (see table 7.1).

Table 7.1
Basic taxonomy of category-based induction

	Single premise	Multiple premises
Specific	(1) Foxes have property S. ――――――― Wolves have property S.	(2) Penguins have property S. Pigeons have property S. Ostriches have property S. ――――――― Sparrows have property S.
General	(3) Robins have property S. ――――――― Birds have property S.	(4) Polar bears have property S. Grizzly bears have property S. ――――――― Bears have property S.

In the following section, we will introduce and briefly discuss the key properties of CBI as identified in empirical research. These phenomena shed light on the conceptual relationships people leverage when evaluating category-based inductive arguments.

7.2.2 Premise-Conclusion Similarity

Not surprisingly, the empirical literature has shown that the strongest criterion used in CBI is the *similarity* among the concepts in the premise and the conclusion (Rips 1975; Osherson et al. 1990; Carey 1985; López et al. 1992). Formally, our expectations about projecting a property from one concept *X* to another *Y* can be modelled as a positive function of their similarity. For instance, arguments like "Ostriches are *S*, then emus are *S*" are seen as stronger than arguments like "Ostriches are S, then blue jays are S," since *sim*(OSTRICH & EMU) > *sim*(OSTRICH & BLUEJAY), where *sim*(*X*, *Y*) denotes a measure of the similarity between concepts *X* and *Y*.

7.2.3 Concept Typicality

Similarity is effective for comparing object concepts at the same hierarchical level, but it falls short when generalizing properties across different levels. In such cases, the typicality of concepts emerges as the primary guide for category-based induction.[2]

The *typicality effect* is the finding that experimental subjects respond more quickly to typical exemplars of an object concept (see section 1.5.2).[3] As it was mentioned earlier, most object concepts exhibit a graded structure (Barsalou 1987; Decock and Douven 2014), meaning that some exemplars represent the category better than others (that is, they are closer to the prototype). For instance, cows are generally seen as very typical instances of the MAMMAL concept, while mice are moderately typical, and whales are highly atypical members.[4]

As in expectation-based reasoning, concept typicality plays a crucial role in CBI. The most robust effect found in the experimental literature is that expectations of property projection are a positive function of premise-typicality. For instance, the inference "Robins have enzyme E; thus, ostriches have enzyme E" is often judged as stronger than "Penguins have enzyme E; thus, ostriches have enzyme E" since robins are more prototypical birds than penguins. To a lesser extent, conclusion-typicality also seems to play a role in CBI. Hampton and Cannon (2004) have shown that arguments with typical concepts in the conclusion (OSTRICH → ROBIN are judged as stronger than arguments with non-typical concepts (CHICKEN → VULTURE).

Furthermore, the typicality effect produces what is called *asymmetry*—that is, the fact that switching the categories from the premises and conclusion often changes the expectations of property projection according to the degree of typicality of the concept in the premise(s). For instance, arguments like "Cows have enzyme E; thus, otters have enzyme E" are considered stronger than arguments like "Otters have enzyme E; thus, cows have enzyme E" since cows are more typical mammals than otters.

7.2.4 Conclusion Homogeneity and Premise Diversity

Another important aspect of CBI is that inferences of argument strength presuppose a shared superordinate concept for the premises. The superordinate concept may be stated explicitly in the conclusion, as in general arguments, or it may remain implicit. For example, in the arguments presented in table 7.1, the implicit superordinate concept in (1) is MAMMAL, while in (2), it is BIRD. In (3) and (4), the superordinate concept appears explicitly in the conclusion. Experimental studies have identified four key phenomena associated with such evoked superordinate concepts: *homogeneity, monotonicity, nonmonotonicity, and premise diversity*.

Homogeneity refers to the inverse relationship between argument strength and the level of abstraction of the object concept in the conclusion: The more abstract and less homogeneous the concepts, the weaker the argument. For example, arguments like "Robins are *S* and blue jays are *S*; thus, all birds are *S*" are judged stronger than "Robins are *S*, and blue jays are *S*; thus, all animals are *S*." This is unsurprising. As we said before, we deal with different degrees of uncertainty when we make or evaluate inductive arguments. The more abstract the object concept in the conclusion, the more information we need from the premises to "cover" it. Hence, category-based inductions with abstract concepts (like ANIMAL or LIVING BEING) involve higher degrees of uncertainty and are harder to support with available premises.

Studies on concept usage, particularly in the prototype tradition, shed light on the homogeneity phenomenon. Concepts vary in their degree of generality, which correlates with the computational cost of their use. Basic-level concepts are cognitively preferred due to their efficiency and centrality in numerous cognitive tasks (Mervis and Rosch 1981). Inductive inference appears to follow a similar principle. Sloman and Lagnado (2005, 106) propose a "preferred level of induction" that aligns with basic-level object concepts, explaining why inductions involving more general categories often seem weaker or less convincing.

Similarity and typicality are key variables in concept usage, potentially explaining the preference for basic-level concepts. These concepts exhibit greater coherence (see chapter 9), which facilitates similarity comparisons among their members. Abstract concepts, by contrast, encompass more diverse and less coherent members, complicating similarity assessments. Furthermore, unlike the concepts basic-level concepts, they lack well-defined prototypes (see Ungerer and Schmid 2006, chapter 2). Consequently, typicality is a more reliable criterion for concept usage at the basic level than for abstract concepts.

Monotonicity refers to strengthening an argument by adding premises, provided their concepts fall within the same evoked superordinate concept (Osherson et al. 1990). For instance, an argument like ROBIN & HAWK → BIRD is weaker than an argument like ROBIN & HAWK & PIGEON → BIRD. However, this strengthening effect does not always hold. Adding premises outside the evoked superordinate concept often weakens the argument. This is called "nonmonotonicity." For example, an argument like ROBIN & HAWK → BIRD will be judged stronger than an argument like ROBIN & HAWK & RABBIT → BIRD. The reason is that introducing RABBIT broadens the contrast class from BIRD to ANIMAL, increasing uncertainty.

Finally, there is *diversity* (Osherson et al. 1990; Feeney and Heit 2011). Empirical studies have shown that diverse object concepts promote expectations regarding property projections. For instance, arguments like "Horses have an ulnar artery and seals have an ulnar artery; thus, all mammals have an ulnar artery" are considered stronger than the argument "Horses have an ulnar artery, and cows have an ulnar artery; thus, all mammals have an ulnar artery." The less similar the object concepts in the premises are the stronger the argument tends to be.

A possible explanation of diversity builds on the idea of "concept coverage" (Osherson et al. 1990). As mentioned before, when evaluating category-based arguments, we evoke a superordinate concept that encompasses all premise concepts. The strength of the argument will depend, to some extent, on how the concepts in the premises *cover* this superordinate concept. For instance, similar concepts such as HORSE and COW have less coverage of the superordinate MAMMAL than dissimilar concepts such as HORSE and SEAL. In this sense, coverage can be described in terms of similarity. We will show how we can model this idea further in section 7.4.1.

Sloman (1993, 253–254) identified an important limitation to the diversity effect: When premises include highly dissimilar concepts, it can actually weaken rather than strengthen the argument. For instance, the argument "German shepherds have sesamoid bones and elephants have sesamoid bones; thus, moles have sesamoid bones" seems stronger than "German shepherds have sesamoid bones and blue whales have sesamoid bones; thus, moles have sesamoid bones." This shows that the inclusion of highly atypical concepts in a premise (BLUE WHALE is a highly atypical mammal), can reduce rather than enhance argument strength.

7.3 Our Model

7.3.1 Inferences and Expectations

CBI has been analyzed from different perspectives, some emphasizing class-inclusion relations (Inhelder and Piaget 1964), others shared features among concepts (Sloman 1998), and others similarity relations (Osherson et al. 1990). However, reasoning about object concepts appears to be a complex mechanism that combines these relations and likely draws on other heuristics as well. In this section, we propose a general model based which offers a natural and relatively simple way to explain concept similarity and typicality, the two primary conceptual relations involved in CBI, while also accommodating other relevant factors.

In our model, we will speak of *expectations of property projection* as what underlies the perceived strength of an inductive argument (see chapter 5). The expression $ExpS(X \rightarrow Y)_Z$ stands for *the expectation that the property S is projected from concept X to concept Y*, with Z as the lowest-level superordinate concept that contains both X and Y.

Let us begin our analysis by examining the simplest case of category-based induction: single premises and specific arguments. For this kind of inductive inference, we want $ExpS(X \rightarrow Y)_Z$ to satisfy the following criteria:

1. It is positively correlated with $sim(X, Y)$.
2. It is positively correlated with $sim(X, p_Z)$, where p_Z is the prototype of Z.
3. It is positively correlated with $sim(Y, p_Z)$.

The rationale for the first condition is that the more similar the categories X and Y, the more expected it will be that Y has property S if X has it. The intuition behind condition (2) is that the closer X is to the prototype of Z, the more expected it is that concept Y has property S, given that X has it. Condition (3) is motivated by Hampton and Cannon's (2004) conclusion-typicality: The more typical Y is for Z, the more expected it is that Y has property S, if X has it.

7.3.2 A Simple Model

To illustrate the basic idea of our approach with a simple model, let us, for the time being, assume that X and Y are represented by small regions, so that we can identify them with points in a conceptual space. Then, given a conceptual space representing categories X, Y, and Z and the distance function d of the space, we can account for the three conditions above by the following equation:

$$ExpS(X \rightarrow Y)_Z = (d(X, Y)\, d(X, p_Z)^a\, d(Y, p_Z)^b)^{-1}. \tag{7.1}$$

Here, a and b are positive constants such that $a > b$. This assumption expresses that premise typicality contributes more to the expectations than conclusion-typicality since, according to the literature, the former is a stronger phenomenon than the latter. The values of both a and b must be empirically determined from data about CBI judgments. We return to this point in section 7.5.

Now, following Shepard's (1987) universal law of generalization, which claims that similarity is an exponentially decreasing function of distance, we can take the logarithm of (7.1) and obtain:

$$logExpS(X \rightarrow Y)_Z = sim(X,Y) + a\, sim(X, p_Z) + b\, sim(Y, p_Z). \tag{7.2}$$

By convention, for any two categories X and Y, $0 \leq sim(X, Y) \leq 1$ and $sim(X, Y) = 1$ if and only if $X = Y$.

Now, equation (7.2) captures the idea that, for single-premise specific arguments, the expectations of projecting S from X to Y are determined by a weighted sum of three factors: premise-conclusion similarity, premise-typicality, and conclusion-typicality.

Equation (7.1), applied to a set of concept prototypes, captures similarity, premise, and conclusion typicality and asymmetry effects in CBI. For instance, when considering the sentence "Emus have property S," people expect more that ostriches also have property S than that penguins have it. This is due to the similarity effect since sim(EMU, OSTRICH) > sim(EMU, PENGUIN). If we construct a "bird space" through some set of prototypes, this inequality would be immediately represented by the relative positions in the space of the two pairs ⟨EMU, OSTRICH⟩ and ⟨EMU, PENGUIN⟩ (see figure 7.1). And it can be measured via the distance function of the space. Since sim(EMU, OSTRICH) > sim(EMU, PENGUIN), it follows from (1) that $ExpS(\text{EMU} \rightarrow \text{OSTRICH})_{\text{BIRD}} > ExpS(\text{EMU} \rightarrow \text{PENGUIN})_{\text{BIRD}}$.

As we mentioned, this model also predicts asymmetry and premise and conclusion-typicality. For instance, $ExpS(\text{ROBIN} \rightarrow \text{EMU})_{\text{BIRD}} > ExpS(\text{EMU} \rightarrow \text{ROBIN})_{\text{BIRD}}$ since $sim(\text{ROBIN}, p_{\text{BIRD}}) > sim(\text{EMU}, p_{\text{BIRD}})$ and $a > b$. Regarding conclusion-typicality, assume, following the bird space in figure 7.1, that sim(OSTRICH, VULTURE) ≈ sim(OSTRICH, ROBIN) and that $sim(\text{OSTRICH}, p_{\text{BIRD}}) \approx sim(\text{VULTURE}, p_{\text{BIRD}})$. Then $ExpS(\text{OSTRICH} \rightarrow \text{ROBIN})_{\text{BIRD}} > ExpS(\text{OSTRICH} \rightarrow \text{VULTURE})_{\text{BIRD}}$ since $sim(\text{ROBIN}, p_{\text{BIRD}})$ is significantly larger than $sim(\text{VULTURE}, p_{\text{BIRD}})$.

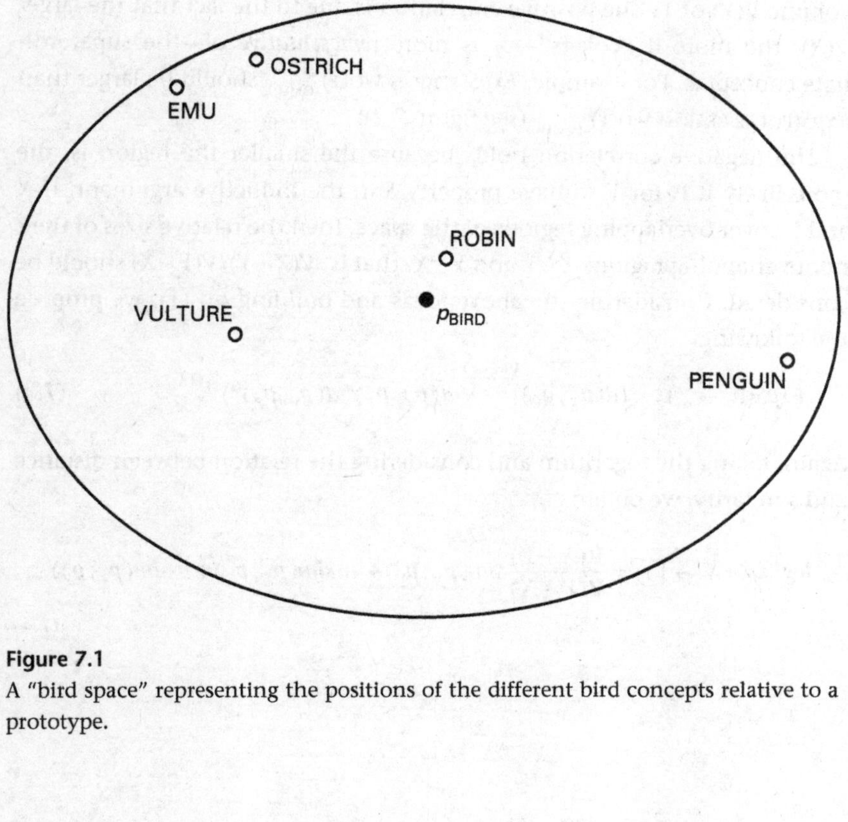

Figure 7.1
A "bird space" representing the positions of the different bird concepts relative to a prototype.

7.3.3 A More General Model

In general, concepts are represented as *regions* of conceptual spaces, not points. To account for this, we now turn to a more general model. Our idea is that the *volumes* of the regions that represent concepts in some conceptual space (areas in the case of a two-dimensional space) can be taken as predictors of expectations, that is, argument strength in CBI.[5] The volume of a concept in a conceptual space depends on the metric assigned to the space, and it is defined in a standard way. It reflects the variability of the property values an object falling under the concept can take in each domain. For instance, it is expected that the concept DOG has a larger volume than the concept TIGER, since dogs can be of several different colors, shapes, and sizes, while tigers have less variability in these domains. The immediate consequence of this is that the more heterogeneous the concept is, the larger its volume will be in a conceptual space.

Coming back to expectations, we assume that $ExpS(X \rightarrow Y)_Z$ is positively correlated with the volume $V(X)$ of X and negatively correlated with the volume $V(Y)$ of Y. The positive correlation is due to the fact that the larger $V(X)$, the more it "covers"—or is more *representative* of—the superordinate concept Z. For example, $ExpS(\text{BEAR} \rightarrow \text{WOLF})_{\text{MAMMAL}}$ should be larger than $ExpS(\text{POLAR BEAR} \rightarrow \text{WOLF})_{\text{MAMMAL}}$ (see figure 7.2).

The negative correlation holds because the smaller the region is, the more likely it is for Y to have property S in the inductive argument. If X and Y cover overlapping regions of the space, then the relative sizes of their nonoverlapping regions $X-Y$ and $Y-X$, that is, $V(X-Y)/V(Y-X)$ should be considered. Considering the above ideas and building on (1), we propose the following:

$$ExpS(X \rightarrow Y)_Z = \left(d(p_X, p_Y)^{\frac{V(X-Y)}{V(Y-X)}} \cdot d(p_X, p_Z)^a \cdot d(p_Y, p_Z)^b\right)^{-1}. \tag{7.3}$$

Again, taking the logarithm and considering the relation between distance and similarity, we obtain

$$logExpS(X \rightarrow Y)_Z = \frac{V(X-Y)}{V(Y-X)} sim(p_X, p_Y) + a \cdot sim(p_X, p_Z) + b \cdot sim(p_Y, p_Z). \tag{7.4}$$

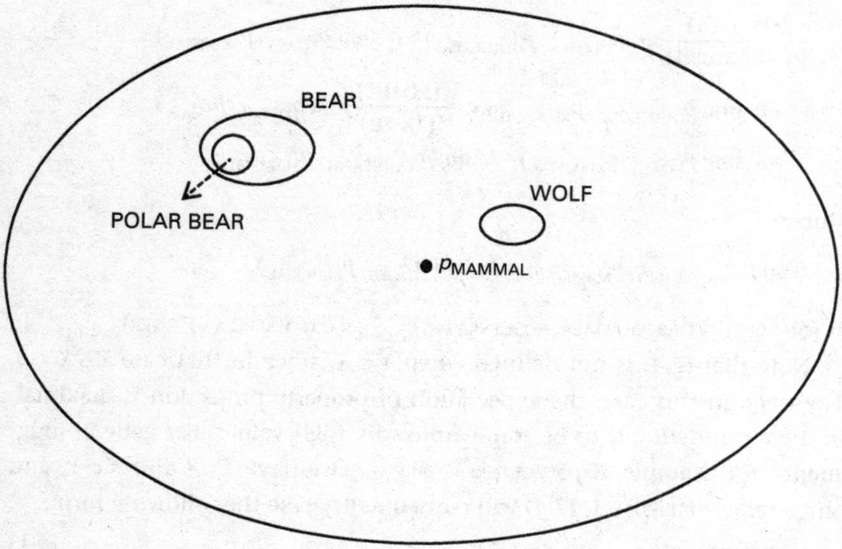

Figure 7.2
A "mammal space" representing the difference in volumes of BEAR, POLAR BEAR, and WOLF.

In cases when X and Y are disjoint regions, which are the most typical ones, the quotient reduces to $V(X)/V(Y)$. And in cases when X and Y are represented by small nonoverlapping regions, we can take $V(X)/V(Y) = 1$ and then equations (7.3) and (7.4), respectively, will have (7.1) and (7.2) as limiting cases.

Just as with (7.1), this new equation predicts premise-similarity, premise and conclusion-typicality, and asymmetry. To see an example of how it works, consider the conclusion-typicality effect. As mentioned in section 7.2.3, some experiments show a robust effect of conclusion typicality in CBI (Hampton and Cannon 2004). For instance, an argument with the categories KOALA → GUINEA PIG should be seen as weaker than an argument like KOALA → TIGER since TIGER is a more typical mammal than GUINEA PIG. Assume, for the sake of argument, that *sim*(KOALA, GUINEA PIG) ≈ *sim*(KOALA, TIGER) and that *V*(GUINEA PIG) ≈ *V*(TIGER). Then, using (7.4), we will have that:

$$\frac{V(\text{KOALA})}{V(\text{GUINEAPIG})} sim(p_{\text{KOALA}}, p_{\text{GUINEAPIG}}) + a\ sim(p_{\text{KOALA}}, p_{\text{MAMMAL}})$$
$$+ b\ sim(p_{\text{GUINEAPIG}}, p_{\text{MAMMAL}}) < \frac{V(\text{KOALA})}{V(\text{TIGER})} sim(p_{\text{KOALA}}, p_{\text{TIGER}})$$
$$+ a\ sim(p_{\text{KOALA}}, p_{\text{MAMMAL}}) + b\ sim(p_{\text{GUINEAPIG}}, p_{\text{MAMMAL}});$$

since

$$b\ sim(p_{\text{GUINEAPIG}}, p_{\text{MAMMAL}}) < b\ sim(p_{\text{TIGER}}, p_{\text{MAMMAL}}).$$

It follows that $ExpS(\text{KOALA} \rightarrow \text{GUINEA PIG})_{\text{MAMMAL}} < ExpS(\text{KOALA} \rightarrow \text{TIGER})_{\text{MAMMAL}}$.[6]

Note that (7.4) is not defined when $Y \subset X$, since in that case $Y - X = \emptyset$. However, in this case, the expectation of property projection is maximal, and we can define it to be some arbitrarily high value.[7] For general judgments, for example, $ExpS(\text{VULTURE} \rightarrow \text{BIRD})_{\text{BIRD}}$, we have $Y = Z$ and $X \subset Y$, and since $sim(\text{BIRD}, \text{BIRD}) = 1$, (7.4) will consequently take the following form:

$$logExpS(X \rightarrow Y)_Z = a\ sim\ (p_X, p_Z) + b. \qquad (7.5)$$

This coheres with the idea that single-premise/general arguments depend mostly on premise-typicality relations—that is, on the idea that subjects

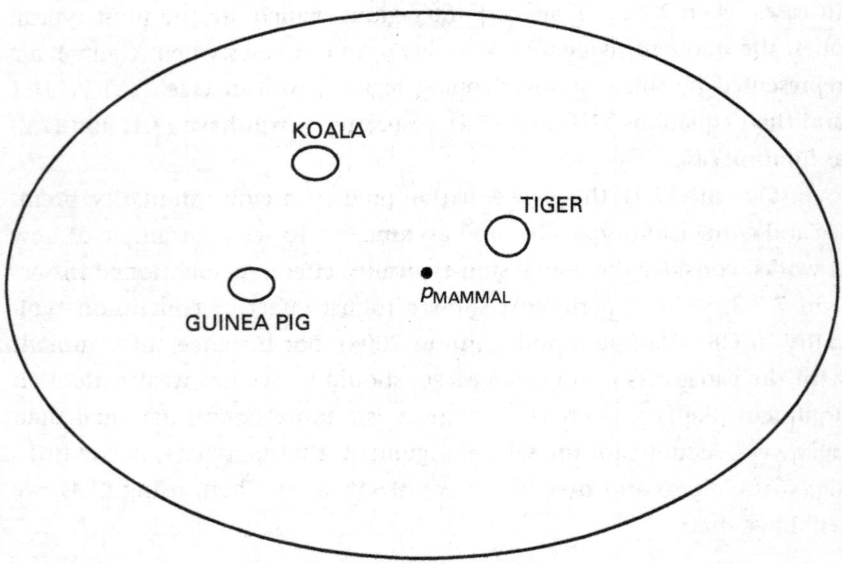

Figure 7.3
A mammal space for categories KOALA, TIGER, and GUINEA PIG.

represent a concept with a certain degree of typicality in the context of a more abstract superordinate concept. This is not a minor point.

In our model, we assume that subjects cannot directly compare concepts from different conceptual levels in terms of similarity (like comparing COLLIE with MAMMAL). Instead, the conceptual relation that works in these cases is typicality, which comes by default for all concepts in a conceptual space, given that conceptual spaces are constructed and articulated around prototypes. As we explain later, we consider this an advantage over the two classical models of CBI, which have more difficulty representing typicality relations among categories.

Now, Sloman (1993) observes that subjects exhibit an "inclusion fallacy" (see section 5.6.2) since the argument "Robins have property S; thus, birds have property S" is judged to be stronger than "Robins have property S; thus, ostriches have property S" even though ostriches form a subset of birds. Our model can explain this phenomenon. To see how, note that since $V(\text{ROBIN} - \text{BIRD}) = \varnothing$, we have that

$$logExpS(\text{ROBIN} \rightarrow \text{BIRD})_{\text{BIRD}} = a\ sim\ (p_{\text{ROBIN}}, p_{\text{BIRD}}) + b,$$

and that

$$logExpS(\text{ROBIN} \rightarrow \text{OSTRICH})_{\text{BIRD}} = \frac{V(\text{ROBIN})}{V(\text{OSTRICH})} sim(p_{\text{ROBIN}}, p_{\text{BIRD}})$$
$$+ a \cdot sim(p_{\text{ROBIN}}, p_{\text{BIRD}}) + b \cdot sim(p_{\text{OSTRICH}}, p_{\text{BIRD}});$$

Then,

$$logExpS(\text{ROBIN} \rightarrow \text{BIRD})_{\text{BIRD}} > logExpS(\text{ROBIN} \rightarrow \text{OSTRICH})_{\text{BIRD}},$$

as long as

$$b(1 - sim(p_{\text{OSTRICH}}, p_{\text{BIRD}})) > \frac{V(\text{ROBIN})}{V(\text{OSTRICH})} sim(p_{\text{ROBIN}}, p_{\text{OSTRICH}});$$

which would typically be the case since $sim(p_{\text{OSTRICH}}, p_{\text{BIRD}})$ is small and $V(\text{ROBIN})$ is comparable to $V(\text{OSTRICH})$. This shows that our model, unlike the similarity-coverage model (Osherson et al. 1990), also predicts results that are not valid under all conditions, but only under certain circumstances.

Our model can also predict the conclusion-specificity phenomenon (Osherson et al. 1990, 187), which says that people tend to judge arguments with more specific concepts as stronger than arguments with more general concepts. For instance, an argument like CROW → BIRD will be judged stronger than the form CROW → ANIMAL. Our model easily explains this because the more general the conclusion concept is, the bigger its volume in the conceptual space, and the further the prototype of this concept will be from the prototype of the premise-concept (see figure 7.4). Considering the above example, we have that

$$ExpS(\text{CROW} \rightarrow \text{BIRD})_{\text{BIRD}} = a\ sim(p_{\text{CROW}}, p_{\text{BIRD}}) + b,$$

and that

$$ExpS(\text{CROW} \rightarrow \text{ANIMAL})_{\text{ANIMAL}} = a\ sim(p_{\text{CROW}}, p_{\text{ANIMAL}}) + b.$$

Then, since $d(p_{\text{CROW}}, p_{\text{ANIMAL}}) > d(p_{\text{CROW}}, p_{\text{BIRD}})$, we have that $a\ sim(p_{\text{CROW}}, p_{\text{BIRD}})$ is greater than $sim(p_{\text{CROW}}, p_{\text{ANIMAL}})$ and, as a consequence, $ExpS(\text{CROW} \rightarrow \text{BIRD})_{\text{BIRD}}$ is greater than $ExpS(\text{CROW} \rightarrow \text{ANIMAL})_{\text{ANIMAL}}$.

7.3.4 Arguments with Multiple Premises

In single-premise arguments, the focus is on the relation between the premise and the conclusion, but when dealing with multiple premises, we must also

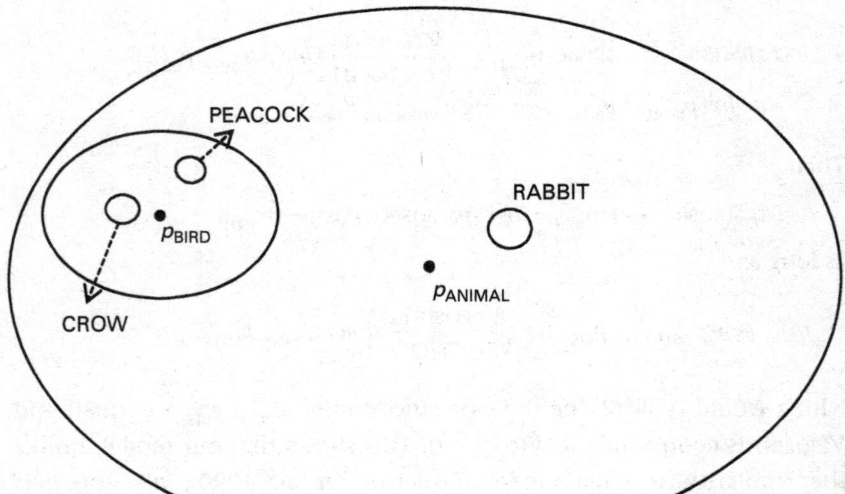

Figure 7.4
An animal space including the subspace BIRD.

account for premise-premise conceptual relations. The main phenomenon to model in these cases is *diversity*: The greater the difference between the concepts in the premises, the stronger the argument. For instance, the argument "Jaguars and leopards have property *P*; thus, otters have property *P*" is weaker than the argument "Jaguars and elephants have property *P*; thus, otters have property *P*," since the concepts JAGUAR and LEOPARD provide less "coverage" of the superordinate concept MAMMAL than JAGUAR and ELEPHANT.

The diversity phenomenon suggests that argument strength is a negative function of premise-premise similarity. One way of modeling this would be to compute pairwise similarities among premise concepts, but this will quicky increase the computational complexity of the process, particularly for arguments with more than two premises. Our proposal tries to avoid this by considering all the concepts in the premises as part of a single inclusive set. In this view, these concepts can be seen as "exemplars" of the evoked superordinate concept. More precisely, we can model n-premises arguments by considering the *convex hull* of the concepts $X_1, X_2, \ldots, X_n$ in the premises. A convex hull of a set S—denoted by $C(S)$—is the smallest convex region containing all elements in S (see Devadoss and O'Rourke [2011] for a detailed explanation).

Convex hulls are also convex regions of n-dimensional spaces with the same geometrical properties as regions in conceptual spaces. The size of their volumes is positively correlated with the number of convex regions they include and the distances among these regions. For instance, in a conceptual space in which all the concepts have similar volumes, the volume of the hull of two contiguous regions is going to be smaller than the volume of two noncontiguous regions of the space.[8] This is precisely the kind of property of interest to represent the diversity phenomena. For example, if we consider the argument described above, in an animal space, the concepts JAGUAR and LEOPARD would be represented by contiguous (or very close) regions in the space, while JAGUAR and ELEPHANT would be far from each other. As a consequence, the volume of C(JAGUAR ∪ LEOPARD) would be smaller than the volume of C(JAGUAR ∪ ELEPHANT), and then it would provide less coverage of the MAMMAL concept (see figure 7.5).

However, one problem with this approach is that we no longer have a natural prototype—like P_X in (7.3)—for the premise. Our solution is to consider an "artificial" prototype p_C, at the centroid of the convex hull

$C(X_1 \cup X_2, \ldots, \cup X_n)$.[9] For convex hulls, we can then reformulate (7.4) for multiple premises in the following way:

$$logExpS(X_1,\ldots,X_n \rightarrow Y)_Z = \frac{V(C(X_1 \cup \ldots \cup X_n) - Y)}{V(Y - C(X_1 \cup \ldots \cup X_n))} sim(p_C, p_Y) + a\ sim(p_C, p_Z) + b\ sim(p_Y, p_Z). \tag{7.6}$$

To see how this formula predicts diversity, consider the example at the beginning of this section. According to (7.6),

$$ExpS(\text{JAGUAR \& LEOPARD} \rightarrow \text{OTTER})_{\text{MAMMAL}} = \frac{V(C(\text{JAGUAR} \cup \text{LEOPARD}))}{V(\text{OTTER})} \times sim(p_C, p_{\text{OTTER}}) + a\ sim(p_C, p_{\text{MAMMAL}}) + b\ sim(p_{\text{OTTER}}, p_{\text{MAMMAL}}).$$

And this is smaller than

$$ExpS(\text{JAGUAR \& ELEPHANT} \rightarrow \text{OTTER})_{\text{MAMMAL}} = \frac{V(C(\text{JAGUAR} \cup \text{ELEPHANT}))}{V(\text{OTTER})} \times sim(p_{C^*}, p_{\text{OTTER}}) + a\ sim(p_{C^*}, p_{\text{MAMMAL}}) + b\ sim(p_{\text{OTTER}}, p_{\text{MAMMAL}}),$$

because $V(C(\text{JAGUAR} \cup \text{ELEPHANT})) > V(C(\text{JAGUAR} \cup \text{LEOPARD}))$, which makes

$$\frac{V(C(\text{JAGUAR} \cup \text{ELEPHANT}))}{V(\text{OTTER})} sim(p_{C^*}, p_{\text{OTTER}}) > \frac{V(C(\text{JAGUAR} \cup \text{LEOPARD}))}{V(\text{OTTER})} sim(p_C, p_{\text{OTTER}})$$

when $sim(p_{C^*}, p_{\text{OTTER}}) \geq sim(p_C, p_{\text{OTTER}})$.

The result also follows in the case in which $sim(p_{C^*}, p_{\text{OTTER}}) < sim(p_C, p_{\text{OTTER}})$, when the difference between $V(C(\text{JAGUAR} \cup \text{ELEPHANT})) / V(C(\text{OTTER}))$ and $V(C(\text{JAGUAR} \cup \text{LEOPARD})) / V(C(\text{OTTER}))$ is enough to make

$$\frac{V(C(\text{JAGUAR} \cup \text{ELEPHANT})}{V(\text{OTTER})} sim(p_{C^*}, p_{\text{OTTER}}) > \frac{V(C(\text{JAGUAR} \cup \text{LEOPARD})}{V(\text{OTTER})} sim(p_C, p_{\text{OTTER}}).$$

Again, this is a conclusion that is not always valid but depends on the relations between the concepts involved.

If Y is a subregion of $C(X_1 \cup X_2 \cup \ldots \cup X_n)$, then $Y - C(X_1 \cup X_2 \cup \ldots \cup X_n) = \varnothing$ so (7.6) is undefined. We can set this to some maximal value for the same reasons as before. For example, if BUZZARD belongs to the convex hull of EAGLE, KITE, and HARRIER, it follows that *Exp*(EAGLE & KITE & HARRIER → BUZZARD)$_{\text{BIRD}}$ would be maximal. This is the first example of a new prediction that emerges from our model. It is an interesting empirical problem, whether this would correspond to the judgment of real subjects. As far as we are aware, this phenomenon has not been tested.

If equation (7.6) is applied to multiple-premise *general* arguments, then $C(X_1, X_2, \ldots, X_n) - Y = \varnothing$, since $C(X_1, X_2, \ldots, X_n) \subset Y$. Then, (7.6) reduces to

$$logExpS(X_1, \ldots, X_n \rightarrow Y)_Z = a\ sim(p_C, p_Z) + b\ sim(p_Y, p_Z). \tag{7.7}$$

Note that when $Y = Z$, $sim(p_Y, p_Z) = 1$ and (7.7) reduces to

$$logExpS(X_1, \ldots, X_n \rightarrow Y)_Z = a\ sim(p_C, p_Z) + b. \tag{7.7′}$$

A problem with this expression is that it does not account for the diversity of $X_1, X_2, \ldots, X_n$, but only the prototype P_c. One way of solving this problem is to let the constant a depend on the proportion of Z that is covered by $X_1, X_2, \ldots, X_n$—that is, $V(C(X_1, X_2, \ldots, X_n))/V(z)$. However, since empirical evidence for this case seems lacking, we will not pursue this theme here.

Let us now see how this model can deal with monotonicity. As mentioned before, the monotonicity effect states that adding premises to a CBI argument increases expectations of property projection when the new premise-concepts are included in the original evoked superordinate concept of the argument. For instance, adding a premise with the concept PIG to the argument (FOX & WOLF) → MAMMAL will strengthen it. Note that adding a premise-concept to an argument will increase the volume of the convex hull of the premises. And in most cases, the volume of that set is negatively correlated to the distance between p_C and p_Y—that is, the more $V(C(X_1, X_2, \ldots, X_n))$ approximates $V(\text{Y})$, the closer p_C is to p_Y. Then, for the above arguments, we have that if p_C is the centroid of C(FOX ∪ WOLF) and p_{C^*} is the centroid of C(FOX ∪ WOLF ∪ PIG) since $V(C(\text{FOX} \cup \text{WOLF} \cup \text{PIG})) > V(C(\text{FOX} \cup \text{WOLF}))$, then $sim(p_{C^*}, p_{\text{MAMMAL}})$ is greater than

$sim(p_C, p_{\text{MAMMAL}})$, and as a consequence *ExpS*(FOX & WOLF & PIG → MAMMAL) is greater than *ExpS*(FOX & WOLF → MAMMAL)$_{\text{MAMMAL}}$.

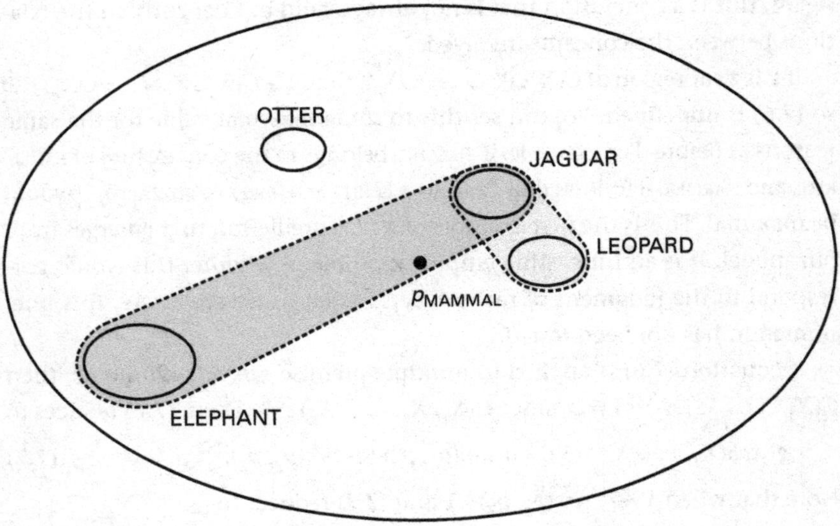

Figure 7.5
A mammal space illustrating that the volume of (ELEPHANT ∪ JAGUAR) is larger than the volume of (JAGUAR ∪ LEOPARD).

This model can also predict Sloman's (1993) observation that diversity has a limit. To analyze his example (mentioned in section 7.2.4), note that $sim(p_C, p_{MOLE})$, where p_C is the prototype of C(GERMAN SHEPARD ∪ ELEPHANT) is considerably larger than $sim(p_{C^*}, p_{MOLE})$, where p_{C^*} is the prototype of C(GERMAN SHEPARD ∪ BLUE WHALE). Similarly, $sim(p_C, p_{MAMMAL}) > sim(p_{C^*}, p_{MAMMAL})$. Then it typically follows that

$$\left(\begin{array}{l}\dfrac{V(C(\text{GERMAN SHEPHERD} \cup \text{ELEPHANT}))}{V(\text{MOLE})} sim(p_C, p_{\text{MOLE}}) \\ +a\ sim(p_C, p_{\text{MAMMAL}}) + b\ sim(p_{\text{MOLE}}, p_{\text{MAMMAL}})\end{array}\right)$$
$$> \left(\begin{array}{l}\dfrac{V(C(\text{GERMAN SHEPHERD} \cup \text{BLUE WHALE}))}{V(\text{MOLE})} sim(p_{C^*}, p_{\text{MOLE}}) \\ +a\ sim(p_{C^*}, p_{\text{MAMMAL}}) + b\ sim(p_{\text{MOLE}}, p_{\text{MAMMAL}})\end{array}\right).$$

Next, suppose that we add to some premise-set a new premise with a concept that is not included in Z. What will happen is that the new modified argument will have a different (and more abstract) evoked superordinate concept Z^* such that $Y \subset Z^*$ with $Z \subset Z^*$. According to the empirical literature, subjects should perceive the new argument as weaker than the original one, making CBI *nonmonotonic*.

Remember the example of nonmonotonicity that we gave in section 7.2.4: The argument (PEACOCK & CROW) → BIRD is stronger than the argument (PEACOCK & CROW & RABBIT) → BIRD (see figure 7.4). According to our analysis, adding the premise RABBIT will change the evoked superordinate concept (Z) from BIRD to ANIMAL. Then, our model correctly predicts that

$$logEXPS(\text{PEACOCK \& CROW} \rightarrow \text{BIRD})_{\text{BIRD}} > logExpS(\text{PEACOCK \& CROW \& RABBIT} \rightarrow \text{BIRD})_{\text{ANIMAL}},$$

since

$$a\ sim(p_C, p_{\text{BIRD}}) + b > a\ sim(p_{C^*}, p_{\text{BIRD}}) + b\ sim(p_{\text{BIRD}}, p_{\text{ANIMAL}}),$$

because $a\ sim(p_C, p_{\text{BIRD}}) > a\ sim(p_C, p_{\text{BIRD}})$ and $b > b\ sim(p_{\text{BIRD}}, p_{\text{ANIMAL}})$.

7.3.5 Knowledge Effects and Nonblank Properties

Until now, our model has considered similarity and typicality as the only relevant factors underlying CBI. However, recent experimental findings show that there may be other explanatory variables that contribute to the dynamics of inductive reasoning (e.g., Rehder 2006; Shafto et al. 2007; Coley and Vasilyeva 2010). For example, there is evidence that knowledge about thematic relations among the concepts involved in the arguments (Coley et al. 2005), as well as expertise in some domain related to the arguments' topic (Proffitt et al. 2000), can play an important role in the agent's expectations of property projection. Furthermore, in most cases of CBI, people use knowledge about the property projected to estimate the argument's strength. This indicates that a complete model of CBI should account for background knowledge effects and consider nonblank properties. We believe the conceptual spaces approach is sufficiently robust to address the most studied knowledge effects of CBI related to non-blank properties. In what follows, we briefly explain how our model can be developed in this direction.

An influential criticism of the similarity-based models of CBI was presented by Heit and Rubinstein (1994). They showed that it is not possible to account for some knowledge effects that influence inductive inferences using only a single similarity measure. In particular, they showed that when an argument involve a nonblank property S, the agents' knowledge of S could modulate the similarity measure used for comparing the concepts in the premise and the conclusion. For example, they showed that an argument of the form CHICKEN → HAWK will be judged stronger than an argument of the form TIGER → HAWK when *S* is an *anatomical property* (such as *has a liver with two chambers*), but the opposite will hold if *S* is a *behavioral property* (such as *prefer to feed at night*). To explain this phenomenon, we extend our model with a similarity measure that assigns greater weight to the dimensions associated with the projected property. When the subject has little knowledge of this property (which is, by definition, the case for a blank property), she will compare categories using a general similarity measure. However, if she has deeper knowledge of *S* (at least about what kind of property *S* is), she is expected to use a similarity measure that gives greater weight to the dimensions related to *S*.

Formally, this can be done by using a weighted distance function like the one introduced in section 7.3.3. Equation (7.1) would be reformulated in the following way:

$$ExpS(X \rightarrow Y)_Z = d^{(S)}(d(X, Y)\, d^{(S)}(X, p_Z)^a\, d^{(S)}(Y, p_Z)^b)^{-1},$$

where the function $d^{(S)}(X,Y)$ is defined as the distance between *X* and *Y* when the dimensions relative to *S* are salient. In the example from Heit and Rubinstein (1994) above, when the projected concept refers to anatomical properties ("has a liver with two chambers"), the model will predict that the argument CHICKEN → HAWK will be judged as stronger than TIGER → HAWK. On the other hand, if the projected concept refers to behavioral properties ("prefer to feed at night"), then the model will predict the converse relation—since now the weight to the behavioral domain will make TIGER more similar to HAWK, just as it was found in the experiments by Heit and Rubinstein.

Medin et al. (2003) showed that property effects also appear in arguments with blank properties. In particular, they discovered a *non-diversity effect by property reinforcement* that occurs when some salient feature shared between the premise-categories leads the agent to produce hypotheses about the nature of the projected property *S* (Shafto et al. 2007). As a result, the agent will use a weighted similarity measure that can override normal diversity effects. For instance, according to what we have seen so far, the argument "Polar bears and antelopes have property *S*; thus, all animals have property *S*" should be considered weaker than the argument "Polar bears and penguins have property *S*; thus, all animals have property *S*," since the first premise set is less diverse than the second. However, in this case, the fact that both polar bears and penguins inhabit cold areas leads agents to hypothesize that property *S* is related to this shared feature. That will weaken the argument, since properties of this kind are atypical regarding animals in general.

Our model handles this naturally. The hypothesis that the property *S* concerns 'coldadapted properties' induces a weighted distance function $d^{(S)}$ (with similarity $sim^{(S)}$) that restructures the space so that differences along dimensions relevant to 'cold adaptation' (e.g., climate/habitat temperature, insulation/thermoregulation) contribute more to distance than irrelevant dimensions. Let p_C and p_{C*} be the centroids of *C*(POLAR BEAR & ANTELOPE) and *C*(POLAR BEAR & PENGUIN) respectively; then, given that for both arguments ANIMAL is the evoked superordinate concept, we use equation (7.7′) and get that

$$logExpS(\text{POLAR BEAR \& ANTELOPE} \rightarrow \text{ANIMAL})_{\text{ANIMAL}} >$$
$$logExpS(\text{POLAR BEAR \& PENGUIN} \rightarrow \text{ANIMAL})_{\text{ANIMAL}}$$

since $a\ sim(p_C, p_{\text{ANIMAL}}) + b > a\ sim^{(S)}(p_{C*}, p_{\text{ANIMAL}}) + b$, because under $d^{(S)}$ p_{C*} will be located in a peripheral, atypical region of animal compared to p_C, lowering $sim^{(S)}(p_{C*}, p_{\text{ANIMAL}})$.

Finally, similar ideas can be applied to explain some of the effects that *expertise* can have in CBI (Proffitt et al. 2000). For a non-expert, the two inferences "Dutch elms have disease A; thus, gingko trees have disease A" and "River birches have disease A; thus, gingko trees have disease A" would, for lack of knowledge, be judged to be equally strong. For a tree expert, however, the knowledge that gingko trees are more similar to Dutch elms regarding which diseases affect them would make the first inference stronger than the second. In brief, for experts, the distance measure $d^{(S)}$ in the model would depend on how S relates to diseases, while this would not affect the non-experts' judgments.

These examples of how knowledge effects can be handled by our model show that it can cover a wide variety of experimental findings from the literature.[10]

7.4 Comparisons with Previous Models

7.4.1 The Similarity-Coverage Model

The first formal model of category-based induction was the similarity-coverage model (SCM), proposed by Osherson et al. (1990). This model evaluates the strength of CBI arguments based on two key factors:

premise-conclusion similarity and the degree of coverage that the premise concept has regarding the lowest superordinate concept encompassing both the premise and conclusion. For specific arguments, argument strength depends solely on premise-conclusion similarity. When an argument has multiple premises, the model employs a maximum rule that estimates premise-conclusion similarity based on the premise whose concept is most similar to the conclusion's concept. For instance, for an argument like "Horses and bats have property S; thus, cows have property S," argument strength will be determined by *Maxsim*{(HORSE,COW); (BAT,COW)}, which will return *sim*(HORSE,COW).

Coverage is a more complex notion. The model assumes—as we do—that CBI with natural concept involves "evoking" an implicit superordinate concept encompassing all the concepts in the argument. Coverage is then defined as a relation between the premise concepts and this superordinate concept, explained in terms of similarity. More specifically, coverage is an average measure of multiple pairwise similarity judgments that compare the premise's concept with exemplars of the superordinate concept. It is also a negative function of similarity among premises, meaning that higher similarity between premises results in lower coverage.

For instance, consider the following arguments:

(a)	(b)
Horses have sesamoid bones.	Horses have sesamoid bones.
<u>Cows have sesamoid bones.</u>	<u>Rats have sesamoid bones.</u>
Mammals have sesamoid bones.	Mammals have sesamoid bones.

Argument (a) is weaker than (b) because the pair (HORSE, COW) provides less coverage of MAMMALS than the pair (HORSE, RAT). The degree of coverage can be associated with the extension of the set that includes all concepts similar to those of the premises. In (a), that set is relatively small because most concepts similar to HORSE are also similar to COW. In (b), however, that set is larger since most concepts similar to RAT are not in the set of concepts similar to HORSE.

Coverage is also related to typicality. SCM assumes that typical categories are associated with larger sets of similar concepts—at the same conceptual level—than atypical ones. For example, the argument "Horses have property *S*; thus, mammals have property *S*" is stronger than "Bats have property *S*; thus, mammals have property *S*" because the set of mammals similar to horses is larger than the set of mammals similar to bats.

Despite its predictive power and success, SCM has several limitations. One drawback is its lack of a psychologically grounded notion of similarity. The model uses similarity as an abstract concept that can be filled with various specific measures rather than providing a precise definition of similarity relations among concepts. This approach, while flexible, fails to build on some theory of conceptual knowledge that could provide a unified explanation for diverse concept-based cognitive phenomena such as categorization, concept formation, and language learning. Furthermore, as Tenenbaum et al. (2007) noted, SCM lacks a systematic mathematical foundation. This limitation is related to the previous point, as the formal structure of the model was not based on any established formal model of inference or conceptual relations. Instead, it was directly designed to model the properties of category-based induction as described in empirical studies.

While closely related to SCM in many respects, our approach addresses these limitations by grounding both formal and psychological foundations in the general theory of conceptual spaces. Furthermore, our model demonstrates significant versatility, not only accounting for the same range of CBI phenomena as SCM but also predicting results in special cases that SCM cannot address.

7.4.2 The Feature-Based Model

Sloman (1993) proposed an alternative CBI model that challenges SCM. He rejected the assumption that reasoning with concepts requires representing their hierarchical structure, citing inclusion fallacies as evidence against this idea, and instead proposed understanding conceptual relations through feature overlap.

Sloman's feature-based model builds on a connectionist framework in which object concepts are represented as sets of features, described as vectors of real numbers from the [0,1] interval. The model explains the ten patterns in CBI accounted for by SCM and predicts three new ones that are supported by empirical findings. The core principle is that argument strength positively correlates with the degree to which the features of the conclusion vector are captured by the premise vector. For instance, in "All

Xs are S; thus, all Ys are S," concepts X and Y can be represented by two vectors $F(X)$ and $F(Y)$ of feature values. The strength of an inductive argument is determined by the following expression:

$$\frac{F(X) \cdot F(Y)}{|F(Y)|^2},$$

where $F(X) \cdot F(Y)$ denotes the dot product, representing the overlap between the features of X and Y, and $|F(Y)|^2$ is a measure of the magnitude of the vector of the concept in the conclusion.[11]

The feature-based approach builds on more solid foundations than the SCM, since it is developed within a connectionist framework.[12] At first sight, this connectionist background may seem to leave no room for our conceptual spaces approach; however, Gärdenfors (1997) and Lieto et al. (2017) have shown that conceptual spaces are totally compatible with connectionist approaches. Indeed, the main ideas behind Sloman's model are translatable into our framework. Both approaches assume that concepts are represented as collections of properties from multiple domains. The key difference lies in how they measure conceptual relationships: Sloman's model uses feature overlap, while the conceptual space approach employs similarity measures across dimensions. In principle, the feature-overlap measure could be replaced by a similarity measure within a conceptual space that encompasses the dimensions of the feature vector. This compatibility suggests that our approach could potentially incorporate the strengths of Sloman's model while offering a more geometrically intuitive representation of conceptual relationships.

A significant advantage our model offers over the feature-based approach lies in treatment of typicality relations. Unlike Sloman's model, which reduces both typicality and similarity to feature overlap, our approach explicitly accounts for independent typicality effects in both general and specific arguments. Sloman's model can explain typicality effects in general arguments, assuming that typical concepts (like APPLE) share more features with their superordinate concept (FRUIT) than atypical ones. However, it falls short in accounting for premise-typicality effects in specific arguments. For instance, if concepts X and Y have equal feature overlap with Z, but X is more typical, Sloman's model would predict equal strength for arguments $X \rightarrow Z$ and $Y \rightarrow Z$ (Heit 1997, 120).

The conceptual space approach does not have this limitation since it can explicitly represent independent typicality relations both in general and specific arguments. For example, consider two arguments of the form QUINCE → PINEAPPLE and APPLE → PINEAPPLE. The concepts APPLE and QUINCE have the same feature overlap with PINEAPPLE, but since APPLE is a more typical fruit than QUINCE, $sim(p_{\text{APPLE}}, p_{\text{FRUIT}})$ is going to be significantly larger than $sim(p_{\text{QUINCE}}, p_{\text{FRUIT}})$, and consequently $ExpS(\text{APPLE} \rightarrow \text{PINEAPPLE})_{\text{FRUIT}} > ExpS(\text{APPLE} \rightarrow \text{QUINCE})_{\text{FRUIT}}$. This constitutes a second novel prediction of our model, which, to our knowledge, has not yet been empirically tested.

One notable aspect of the conceptual spaces model is that it combines key features of both SMC and the feature-based model: It is a similarity-based approach incorporating a feature-based view of concepts. Additionally, it offers a major theoretical advantage by drawing on the theory of conceptual spaces, which explains how domains are formed, structured, and grounded in perception and action. This grounding connects our formal framework to a comprehensive psychological account of the nature and organization of conceptual systems and also provides a formal model of key cognitive mechanisms involved in conceptual processes, giving our approach a solid mathematical foundation. An additional strength of our model is its ability to account for the conclusion effect observed by Hampton and Cannon (2004), a phenomenon that neither the SCM nor the feature-based model can explain (Feeney 2018, 172–173).

7.4.3 Bayesian Models

Bayesian approaches have recently gained prominence in the literature on CBI. Heit (1998) pioneered this direction with a computational analysis of how knowledge of properties influences CBI. His idea is that, when evaluating a CBI argument, subjects estimate the probability of property projections among concepts based on their assessment of the property's range—that is, the set of concepts for which the property holds or not. This estimation draws on prior knowledge of familiar properties, assuming that the projected property will be distributed similarly.

For instance, in an argument such as "*X* has property *S*; thus, *Y* has property *S*," subjects consider four hypotheses about the range of property *S*: (1) *S* applies to both *X* and *Y*, (2) *S* applies to *X* but not *Y*, (3) *S* applies to *Y* but not *X*, and (4) *S* applies to neither *X* nor *Y*. The prior probability distribution for these hypotheses may vary according to the similarity between *X* and *Y* or other conceptual relations. Then, taking the argument's premise as evidence, subjects update their beliefs about these hypotheses and estimate the probability of the conclusion using Bayes' theorem.

Heit shows that his approach predicts as many properties of CBI as SCM and the feature-based model. However, it has also an important drawback: It doesn't include any mechanism for estimating the prior distribution over hypotheses about a property's range. This problem primarily stems from the Bayesian approach's focus on property relations rather than concept relations. Unlike other models, including our conceptual spaces approach, Heit's model centers on how properties are distributed across concepts instead of how concepts relate to each other.

Tenenbaum et al. (2007) built on Heit's approach to tackle the challenge of estimating prior probabilities. They defined structures representing the agent's knowledge of conceptual relations in different domains and how properties align with them. These structures determine the prior probability of projecting a property *P* from concept *X* to *Y* within the same domain, and these priors can be then updated using Bayesian rules when evaluating specific arguments. In this way, this approach grounds prior probabilities in conceptual knowledge and property-concept interactions, providing a more comprehensive framework for modeling inductive reasoning.

This Bayesian framework can accommodate various knowledge structures, including taxonomic systems, causal structures, and spatial knowledge. This versatility is a significant advantage over the SCM model and Sloman's model, which have serious troubles dealing with forms of inductive reasoning that don't involve natural concepts. There are, however, considerable drawbacks to Bayesian models of CBI. For example, they lack a natural way to represent similarity and typicality—which are crucial aspects of human conceptual reasoning—and computational implementations of probabilistic reasoning are highly resource-intensive. In our opinion, these limitations affect the psychological plausibility of Bayesian models (see Jones and Love [2011] for a general criticism of the use of Bayesian models in cognitive science).

Beyond Bayesian models, the most advanced probabilistic models of reasoning are those provided by large language models based on deep learning. Han et al. (2022) compared the ability of GPT-3 to reason about category-based induction with how humans' reason. Their results show that GPT-3 accounts for some aspects of human reasoning with CBI, but overall, the match is relatively poor. For example, GPT-3 has problems with premise diversity and with premise monotonicity. Han et al. (2022) argue that the problems arise because the inferential processes of GPT-3 are different from those used by humans.

7.5 Experiments Testing the Model

A major challenge faced by researchers working on CBI is developing quantitative tests for the available models. The framework presented in section 7.4 opens up a new methodology for investigating CBI. Its distance measure, together with the similarity and betweenness relations it generates, will allow for new and more precise quantitative predictions. We have already mentioned the prediction that when Y is a subregion of $C(X_1 \cup X_2 \cup \ldots \cup X_n)$, then $ExpS(X_1, X_2 \rightarrow Y)_Z$ should be maximal. For regions Y, X_1 and X_2, say that Y lies *between* X_1 and X_2 if for every y in Y there are points x_1 and x_2 in X_1 and X_2 respectively, such that y is between x_1 and x_2. Given this definition, a special case of the prediction above is that if Y lies between X_1 and X_2, then $ExpS(X_1, X_2 \rightarrow Y)_Z$ should be maximal. Another new prediction concerns explicit representations of independent typicality relations, as was discussed in section 7.5.2.

Some further predictions are related to the introduction of the notion of *volume* of a concept. First, our model predicts that premise-specificity is negatively correlated to argument strength. More formally, it is to be expected that for concepts Y, X_1 and X_2, if $X_1 \subset X_2$ then $ExpS(X_2 \rightarrow Y) > ExpS(X_1 \rightarrow Y)$ because $V(X_2) > V(X_1)$. For instance, arguments of the form GERMAN SHEPHERD → COW (or MAMMAL) should be considered weaker than those of the form DOG → COW (or MAMMAL). Second, our model predicts that for concepts X_1, X_2 and Y, if it's the case that X_1 and X_2 are equally typical, but that $V(X_1) > V(X_2)$ then $ExpS(X_1 \rightarrow Y) > ExpS(X_2 \rightarrow Y)$. These two predictions hold *ceteris paribus*.

To test these predictions, operational procedures for determining betweenness, similarity, and distances are needed. As noted in section 2.3, there are general methods for estimating psychological distances, such as multi-dimensional scaling (MDS) (see Hout et al. [2013] for a review) and principal component analysis (PCA) (Abdi and Williams 2010). For example, by asking subjects to judge the similarities of several different concepts, the data can be analyzed by MDS or PCA to generate a low-dimensional conceptual space with a distance measure. Once the distance is established, similarity and betweenness can be determined, and the predictions presented above can be tested.

As an example of the relevant type of data collection, Hampton and Cannon (2004) asked subjects to rate the premise typicality, conclusion-typicality, and premise–conclusion similarity on a seven-graded scale. This data could also have been used to estimate an underlying distance measure that would have allowed equations (7.1) or (7.2) to be tested.[13]

There is extensive experimental evidence concerning CBI, as documented by Heit (1997) and Feeney (2018). Our model can explain a large majority of these experimental results, as also demonstrated in Osta-Veléz and Gärdenfors (2020). While most experiments in this field report only qualitative results, our model allows for quantitative predictions once the metric of the underlying conceptual space has been estimated.

Douven et al. (2023) present a study that explicitly tests some of these quantitative predictions. Their stimuli were generated from forty-nine pictures of objects resembling cups, vases, and bowls, originally used in Douven (2016). These objects were organized in a 7-by-7 grid along the dimensions of width and height. Importantly, the psychological distances between these objects had been carefully determined in the earlier study, providing a preexisting distance function for the object space. The experiment tested two hypotheses generated from our model: (1) The strength of proximity-influenced arguments can be predicted from the distance in the conceptual space between the object in the premise and the object in the conclusion, as well as from how typical these objects are of the projected property. (2) Truth ratings of conditional sentences mentioning two embodying proximity-influenced inferences can be predicted on the exact same basis.

To test the first hypothesis, the participants were shown twelve pairs of objects randomly selected from the forty-nine in the space. For each pair, the participant was asked to suppose that the vessel that appeared on the left was a vase and then to indicate whether that gave them reason to believe that the vessel on the right was a vase as well. The response had to be given on a seven-point Likert scale. For the second hypothesis, the participants were shown the same twelve pairs. In this part, the participants were asked to evaluate the truth of a conditional sentence of the form "If the vessel on the left is a vase, so is the vessel on the right." They were then asked to indicate whether they thought this was true or false, but they were also given the option "neither."

The results provided strong support for both hypotheses: Antecedent–consequent similarity was strongly associated with perceived argument strength, respectively, and degree of truth. For further details, see Douven et al. (2023).

Another experiment that relates directly to the model presented here is presented by Lu et al. (2023). They use EEG to study the timing of brain activities during the period a subject considers a case of category-based induction. Comparing three different processing models, the best support is found for the "default-interventionist" model, which states that premise diversity and premise typicality are activated sequentially (instead of in parallel). Premise diversity activated brain areas that suggest that premise diversity was facilitated during an early time window and then revised by premise typicality during a later period.

Consistent with our model, the non-diverse premise inductions had a higher proportion of correct responses, greater inductive strength, and shorter reaction times than the diverse premise ones. Diverse premise problems involve more processes of revision or updating, as indicated by higher EEG amplitudes. Lu et al. (2023: 11) also note that their study supports our criticism of Bayesian models of case-based induction. They write, "Bayesian models focus on property relations and ignore how individuals represent concepts. Thus, such models lack psychological reasonability."

7.6 Conclusion

In recent decades, category-based induction has become a significant research focus in the psychology of reasoning. In this chapter we have presented a mathematical model that accounts for much of the available empirical evidence on such inferences. Our model, constructed using similarity and typicality measures, extends beyond previous frameworks. It encompasses Osherson et al.'s (1990) similarity-coverage model and Sloman's (1993) feature-based approach while generating novel predictions. Furthermore, it builds on solid formal foundations, and it relies on a systematic theory of conceptual knowledge that has been applied successfully to the explanation and modeling of several concept-based cognitive phenomena.

8 Analogies

Analogical reasoning is a form of similarity-based inference as fundamental to cognition as inductive reasoning (Hofstadter and Sander 2013). It involves mapping structural similarities between two domains—a source and a target—that are typically regarded as fundamentally different. For instance, arteries can be analogically compared to highways due to their primary function of transportation despite a vast backdrop of dissimilarities among them. Likewise, Rutherford's atomic model drew parallels between the solar system and atomic structure, illustrating the idea of electrons orbiting a nucleus.

The semantic profile of analogical statements differs from that of factual statements, which convey information about states of affairs in the world. Analogies have a primarily *epistemic* goal: They highlight higher-order similarity relations between different semantic domains, helping to organize conceptual knowledge and facilitate the formation of abstract categories (Gentner and Hoyos 2017).

The vast literature on analogy in the cognitive sciences supports this claim. It has been shown that analogy mechanisms play a crucial role in language acquisition (Behrens 2017), concept learning (Tomlinson and Love 2006; Gentner and Hoyos 2017), teaching (Vosniadou 1995), scientific reasoning and discovery (Oppenheimer 1956; Gentner 2002; Nersessian 2010), and AI (Jani and Levine 2000; Barbot et al. 2019; Mitchell 2021).

8.1 Different Kinds of Analogies

There are two types of analogical structures: *direct* and *composed* analogies. The former compares an individual source with an individual target, like in the examples given above. Composed analogies—also called "proportional"—on the other hand, compare two pairs of objects according to some salient relation between the elements in each pair. For instance, the sentence "The foot is to the leg as the hand is to the arm" is a composed analogy since the salient (mereological) relation between FOOT and LEG is "symmetrical" to the mereological relation between HAND and ARM. Within AI and in the cognitive sciences, most attention has been focused on direct analogies. In this chapter, however, we focus on composed analogies. We will use the notation *A:B::C:D* for a composed analogy where the pair *A:B* is compared to the pair *C:D*.[1]

Our main goal is to show that conceptual spaces can provide an explanatory framework for how humans reason about composed analogies. Building on an early geometrical model of analogical reasoning known as the "parallelogram model" (Rumelhart and Abrahamson 1973), we will show that composed analogies can be analyzed in terms of similarity relations across dimensions that are salient to the elements involved in the analogies. We will distinguish between four different types of composed analogies: (1) categorical (dimensional) relations (e.g., TUNA:SHARK or HOT:COLD), (2) taxonomic relations (e.g., LIZARD:REPTILE), (3) event-based relations (e.g., OPEN DOOR:CLOSED DOOR), and (4) part-whole relations (e.g., FOOT:LEG). We outline algorithms for modeling these four types within the framework of conceptual spaces (see also Osta-Vélez and Gärdenfors 2022).

8.2 The General Search Procedure

This chapter introduces a novel approach to composed analogies, treating them as structured *search problems*.[2] The *search space* encompasses all the concepts in a *lexicon L*. The *initial state* consists of an incomplete analogy *A:B::C:X*, where "*X*" is an unknown lexical concept or property. The *goal condition* requires identifying at least one element *X* within *L* that replicates the semantic relation *A:B* in *C:X*. The search algorithm employs a specific heuristic derived from the analysis of the semantic relation in the initial state. The process culminates in a final state in which a lexical concept, or an ordered set of concepts, satisfies the goal condition. The process of solving a particular analogy *A:B::C:X* involves the following steps:

(1) Identifying the semantic relation in *A:B* and restricting the search space accordingly in the light of *C*.

As we will show below, the semantic relation can be of varying types. The outcome of step (1) is a restricted search space. For instance, if the semantic relation is taxonomic, the search space will be restricted to the set of lexical items that are taxonomically related.

(2) Specifying the search algorithm.

We present four search algorithms for different semantic relations. The kind of semantic relations will define the similarity space in which the algorithm works.

(3) Apply the search algorithm to identify one or more lexical concepts that can meet the goal condition.

The following section presents a background to research on analogies in human reasoning. In section 8.3, we discuss the role of similarity in analogies. In section 8.4, we show that the salience of different domains is important in judging analogies. In section 8.5, we discuss analogies whose terms are all concepts at the same level of abstraction (e.g., DOG:WOLF::CAT:LYNX) and introduce the parallelogram model as a basis for a search algorithm. Section 8.6 is devoted to property-based analogies of the type PEAR:GREEN::BANANA:YELLOW, and we present a typicality criterion to be used in the search procedure. For these analogies, an object concept is compared to a characteristic property. In section 8.7, we turn to event-based analogies, exemplified by HORSE:GALLOP::MAN:RUN. For this type, *force* and *result vectors* will be central. The final type of analogy is part-whole analogies of the form HAND:ARM::FOOT:LEG. We argue in section 8.8 that also such mereological relations can be modeled by using conceptual spaces.

Our focus in this chapter is to model human reasoning, but the models we present are amenable to implementations in AI systems via the search algorithms we identify. As we will show for all four types of analogies, using conceptual spaces as a basis for the models opens up new types of computational implementations.

8.3 Analogical Similarity

Similarity is the fundamental relation underpinning analogical processes (Cummings 2014; Holyoak 2012). However, as we have previously noted, similarity itself is a problematic notion and many of its puzzling features permeate the study of analogy (Goodman 1972; Smith 1989). The sentence "Whales are like sharks" illustrates this complexity. Its accuracy depends on the criteria of comparison: It holds for perceptual features but fails when considering biological and taxonomic characteristics. In general, similarity judgments require additional context to specify the relevant aspects of comparison.

What distinguishes analogies from "literal" similarity judgments such as the one above is that the former is based on a relation of "structural" or "relational" similarity between the objects being compared (Ortony 1979; Gentner and Markman 1997). For example, consider the statement "Sharks are to the ocean what lions are to the savannah." Evidently, this statement cannot be interpreted as pointing to a relation of literal similarity between the objects being compared but rather to a functional or conceptual similarity in their respective roles within their environments. This shows that a theory of analogy requires a specification of what kind of similarity relation underlies the process.

The most influential approach to *analogical similarity* is the structural alignment view (SAV), pioneered by Gentner and colleagues (Gentner 1983; Markman and Gentner 2000; Gentner and Holyoak 1997). According to SAV, analogy depends on comparing two representational structures consisting of objects, properties (unary predicates), and high-order relations. The comparison is facilitated by a one-to-one mapping between the elements of these structures, resulting in three sets:

1. *Commonalities*: shared characteristics or objects among the structures.
2. *Alignable differences*: pairs of characteristics or objects that are in principle different but can still be put in correspondence because they are instances of a common superordinate concept.
3. *Non-alignable differences*: elements of some of the structures that fail to map because they fit neither of the previous two sets (e.g., "gills" in "Sharks are to the ocean what lions are to the savannah").

Crucial to this theory is that alignable differences are more salient than non-alignable differences (Gentner and Markman 1997, 50).

A central idea in SAV is that analogical similarity is predominantly *relational*, focusing more on mapping higher-order relations than on matches between objects or attributes (see Gentner et al. 1995). Gentner and some

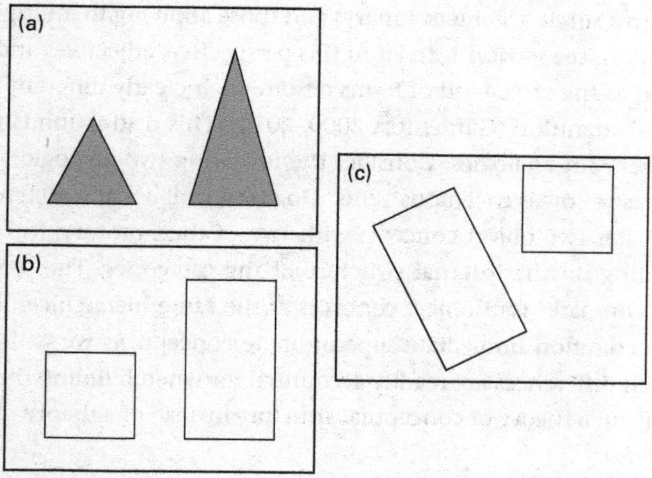

Figure 8.1
Three pairs of analogous objects.

other authors go so far as to claim that our ability to handle relational predicates in analogies marks the difference between human and nonhuman cognition (Gentner 2003). To illustrate this idea, consider the scenes depicted in figure 8.1.

At the object/attribute level, images *b* and *c* are highly similar: They have many common properties and only one alignable difference: the spatial configuration of the objects. On the other hand, *a* and *b* have many alignable differences, like shape and color in every object mapped, but also an important commonality: a similar relational structure (*shorter_object left_of_taller_object*). This last commonality makes *a* and *b* analogies, while *b* and *c* share only a relation of *literal* similarity.

The above example illustrates how analogical similarity is of a special kind and cannot be captured by the intuitive idea of similarity as a function of a positive difference between shared and unshared features (see also Goldstone et al. 1991). SAV succeeds in proving this point as well as in advancing a richer notion of similarity that distinguishes between varieties of differences. Nonetheless, the approach has some limitations. First, SAV does not build on any specific theory of conceptual structure. As in most computer science, SAV instead analyzes analogical representations from a propositional perspective, classifying predicates simply by arity: those applying to a single argument (unary) and those applying to multiple arguments (n-ary) (see section 1.2). From this perspective, adjectives and nouns are treated as the same kind of terms despite their clearly different roles in conceptual cognition (Gärdenfors 2000, 2014). This distinction is particularly relevant for analogies. Consider the following two analogies: "Red is to apples as yellow is to lemons" and "Dog is to wolf as cat is to lynx." The former relates two object concepts with two of their prototypical properties, building on the internal structure of the categories. The second, in contrast, compares four object concepts at the same hierarchical level, all sharing a common immediate superordinate concept. As we will demonstrate, such differences are readily accounted for when building the theory of analogy on a theory of conceptual structure instead of a theory of propositional form.

Second, while relational predicates certainly play a crucial role in analogies, it is problematic to claim that analogical similarity can be fully reduced to *relational* similarity. Analogies like "Sharks are to the ocean what lions are to the savannah" do not build on any relational concept but rather on the salience of shared dimensions. Moreover, most relational predicates can be analyzed in terms of single dimensions.[3] For instance, comparative adjectives like *Taller*(*x*,*y*) or *Younger*(*x*,*y*) express a difference with respect to one shared dimension of the objects (in these cases, *height* and *age* dimensions). In such cases, the crucial factor is not the relational structure but our knowledge of the dimensions being compared.

Third, computational implementations of relational similarity are difficult to build in a systematic manner. The distance functions of conceptual spaces, as used in the criteria for different kinds of analogies presented in this chapter, offer new avenues for implementation. We outline how this can be achieved by describing various search procedures. Our analysis will demonstrate that multiple computational methods, rather than a single approach, are necessary to handle diverse analogies. This approach will be crucial for AI systems aiming to mimic human reasoning, where analogies play a central role.

8.4 The Role of Dimensional Salience

Our approach draws on the last two critical points raised in the previous section. First, we argue that composed analogies need to be analyzed within the framework of a theory of conceptual structure. Second, we propose that analogical similarity primarily depends on *dimensional salience*—specifically on identifying one or more dimensions that serve as a comparative frame for the lexical concepts in the analogy. An analogy's "aptness" or "quality" correlates with the degree of salience these dimensions have for the concepts or properties appearing in the analogy. To illustrate this principle, consider the following analogies:

(8.1) DOG:PUPPY::CAT:KITTEN

(8.2) SWEET:APPLE::SOUR:LEMON

(8.3) HOT:WARM::COLD:COOL

(8.4) RABBIT:LION::TUNA:SHARK

Each of these analogies consists of projecting a salient semantic relation among the concepts of the first pair onto the concepts of the second. This relation hinges on identifying one or more dimensions of the concepts that can serve as *analogy factors*. In (8.1), the analogy factor is the *age* dimension, in (8.2) is *taste*, in (8.3) is the *temperature dimension*, and in (8.4) is *size* and *ferocity*. The analogy factor is typically differential: It selects a dimension in which the concepts of the first pair have significantly different values.

A key challenge in evaluating an analogy is determining which dimensions, among the many that constitute the involved concepts, best support the analogical relation. For instance, *size* is a good candidate for an analogy factor in (8.4), while *color* clearly is not. In our approach, the dimensions that are prioritized as potential analogy factors are the most *salient* of the concepts in the first pair (see section 2.6). A straightforward prediction is that the processing speed of an analogy will correlate positively with the degree of saliency of the analogy factors and negatively with the number of dimensions that can serve as analogy factors. For instance, (8.3) is a simple analogy because a single dimension relates its four terms. In contrast, (8.4) affords multiple dimensions as potential analogy factors and, consequently, is more complex.

While classical approaches tend to seek highly general models of analogy (Gentner 1983; Holyoak and Thagard 1989), our view diverges in treating analogy as concept specific. In other words, analogies draw on the representational structures associated with the word classes they involve. Since different word classes represent distinct types of concepts (Gärdenfors

2000, 2014), a comprehensive theory must integrate several submodels. In the remainder of this chapter, we propose such models.

We focus on nouns, adjectives, verbs, and meronomic relations. While this list is not exhaustive, we believe it sufficiently demonstrates the utility of conceptual spaces in analyzing analogies. Unlike traditional approaches in logic and computer science that treat all predicates uniformly, we aim to show that distinguishing predicates by their conceptual roles will yield more fruitful computational systems, specified as distinct search procedures.

8.5 The Parallelogram Model Generalized

One of the earliest models of analogy was developed by Rumelhart and Abrahamson (1973), who proposed that analogical inference relies on the organization of knowledge in semantic memory. They argued that analogical similarity could be expressed as a function of semantic distance between concepts represented as points in a multidimensional space. Specifically, they showed that analogies of the form *A:B::C:D*, where the elements are object concepts, must adhere to a "parallelogram rule" that states that the vectorial distance between concepts *A* and *B* must be equal (or highly similar) to the vectorial distance between *C* and *D* (see figure 8.2).

In their model, analogical inference is seen as a *choice problem* in which agents first represent the categories *A*, *B*, and *C* as points in a multidimensional space and then select a category *D* that is the best option for fulfilling

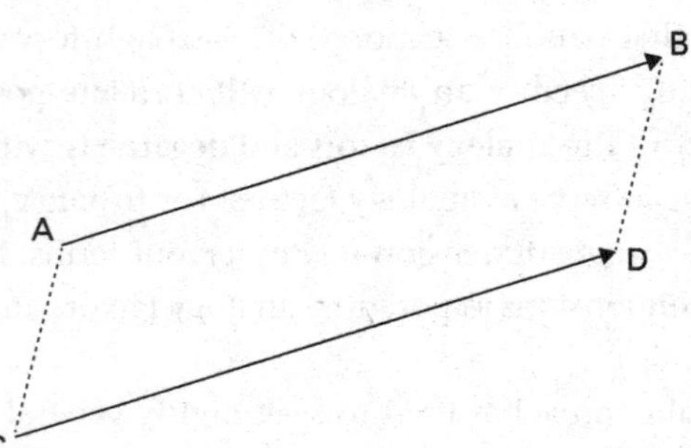

Figure 8.2
Vector relations among the four analogy terms represented as points in a space, according to Rumelhart and Abrahamson (1973).

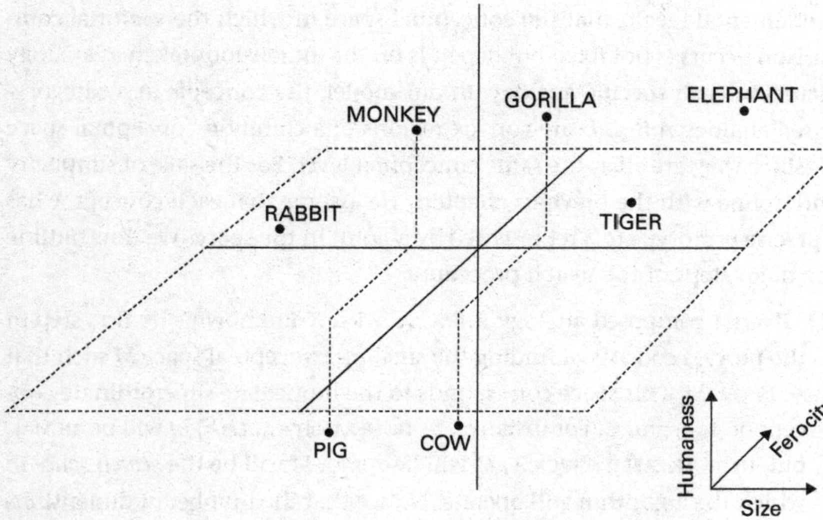

Figure 8.3
A mammal space organized around the size, humanness, and ferocity dimensions, illustrating the example from Rumelhart and Abrahamson (1973, 3).

the parallelogram rule. Roughly, this rule consists of fixed vector addition and subtraction operations: $D=(B-A)+C$.

In a series of experiments using Henley's (1969) three-dimensional mammal space (see figure 8.3), Rumelhart and Abrahamson showed that when presented with analogy problems like MONKEY:PIG::GORILLA:*X*, with RABBIT, TIGER, COW, and ELEPHANT as alternatives for *X*, subjects rank the four options following the parallelogram rule. The parallelogram model predicts that COW is the preferred solution. Their experiments clearly supported the model.

Overall, we believe this model provides valuable insight into category-based analogies. However, it has two significant limitations. First, it lacks explanatory depth, as it neither allows for the identification of analogy factors nor provides a notion of analogical similarity. Second, it can only handle analogies where all terms are at the same hierarchical level. Analogies like SWIM:FISH::FLY:BIRD or BEAR:MAMMAL::LIZARD:REPTILE cannot be analyzed by this model because not all object concepts in them have the same vector representation and, consequently, are incomparable.

We next present a generalized version of the parallelogram model that follows the semi-algorithmic approach previously described. The fundamental idea is that the conceptual space in which the vectorial comparison occurs is not fixed but depends on the dimensions taken as analogy factors in each specific analogy. In our model, the concepts in a category-based analogy *A:B::C:D* are convex regions of a common conceptual space *M*, since they are all at the same conceptual level. For the sake of simplicity and in line with the previous chapters, we assume that each concept *X* has a precise prototype (p_X) represented by a point in the space. We now outline the main steps of the search procedure.

(1) Given a composed analogy *A:B::C:X*, with *X* unknown, the first step in the process consists of finding the smallest conceptual space *M* such that $A, B, C \subset M$. This space corresponds to the immediate superordinate concept of *A*, *B*, and *C*. For instance, in TIGER:RABBIT::EAGLE:*X*, *M* will be ANIMAL, but, in TIGER:RABBIT::TRUCK:*X*, *M* will be THING. *M* will be the *search space* in which the algorithm will operate. Notice that the number of dimensions apt for establishing an analogical comparison depends on the specificity of *M* (i.e., its place in Rosch's [1978] vertical level of categorization). Dimensions that are available for animals in TIGER:RABBIT::EAGLE:*X*, like *diet*, *ferocity*, or *humanness*, cannot be applied to *things* in TIGER:RABBIT::TRUCK:*X*.

(2) The second step consists in selecting from *M* a set of *salient* dimensions $D_1, D_2, \ldots, D_n$ where the salience is generally determined by the difference between *A* and *B* (often, only one dimension is relevant). For instance, consider the relation TIGER:RABBIT::EAGLE:ROBIN. *Ferocity* and *size* are two salient dimensions of the concept ANIMAL, since an important difference can be established between TIGER and RABBIT across these dimensions. Since these differences can be replicated for the concepts EAGLE and ROBIN, the analogy is sound. Choosing these dimensions as a frame of comparison will generate a "new" lower-dimensional conceptual space M^* with a distance function d^*.[4] This modulated distance function computes what we call *analogical similarity*, which constitutes the main difference compared to Rumelhart and Abrahamson's parallelogram model.

(3) The search algorithm's last step is applying the parallelogram rule on M^* for choosing the optimal solution to *x* in *A:B::C:X*. For this, we start with the prototypes p_A, p_B, p_C, and the vector $\overrightarrow{p_A p_B}$ in M^*, and we find the point $y \in M^*$, that is the head of a new vector $\overrightarrow{p_C y}$ that is as close as possible (same direction and magnitude) to $\overrightarrow{p_A p_B}$. The concept $X \subseteq M$ that gives the strongest analogical relation will be the one whose prototype

p_X is closer to y than any other prototype in M, that is, p_X such that $d^*(p_X,y) < d^*(p_Z,y), \forall p_Z \in M$.

Let us illustrate this procedure with a toy example. Consider the incomplete analogy MOUSE:WOLF::RABBIT:X and a reduced search space with concepts HIPPO, BUFFALO, ELEPHANT, and GORILLA. M will be the mammal space used by Rumelhart and Abrahamson (1973) (see figure 8.2), and the dimensions that will serve as the frame of comparison in the space M^* will be *size* and *ferocity*, due to the salient difference that the concepts MOUSE and WOLF maintain across them. The ***humanness*** dimension in figure 8.3 is less salient and will not be part of M^*. Then, in a weighted conceptual space M^*, a point y will be determined as the head of a vector with tail in the prototype of RABBIT that is equivalent to the vector formed by the prototypes of MOUSE and WOLF. Assuming the positions of the prototypes as depicted in figure 8.4, the prototype of BUFFALO is the optimal solution to the analogy since it is closer to y than any other prototype in M^*.

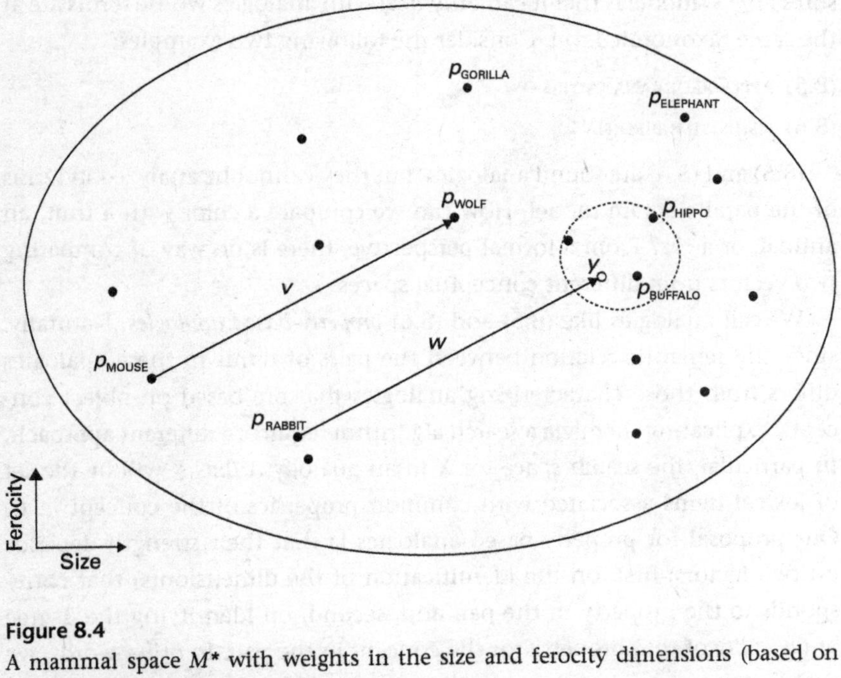

Figure 8.4
A mammal space M^* with weights in the size and ferocity dimensions (based on Rumelhart and Abrahamson 1973).

Analogies are not all or nothing; they have degrees of aptness or soundness. For instance, concepts very close (in the weighted conceptual space) to the optimal choice in a category-based analogy might also be good solutions. In addition to this, it is possible that different sets of dimensions are taken as analogy factors, generating multiple possible sound analogies. We believe that, in most cases, a particularly salient set of dimensions will produce the strongest analogical relation. However, offering a systematic criterion for finding it is rather complicated, for it is strongly dependent on the subjects' knowledge of a particular semantic domain, as well as on the semantic intuitions rooted in a community of speakers. Ultimately, finding the set of salient dimensions for a given concept is an empirical question.

8.6 Property-Based Analogies

8.6.1 From Object Concepts to Properties

As mentioned earlier, an important limitation of Rumelhart and Abrahamson's (1973) model is that it can only deal with analogies whose terms are at the same taxonomic level. Consider the following two examples:

(8.5) APPLE:RED::BANANA:YELLOW

(8.6) FISH:SWIM::BIRD:FLY

(8.5) and (8.6) are sound analogies, but they cannot be analyzed in terms of the parallelogram model. How can we compare a color with a fruit, an animal, or a car? From a formal perspective, there is no way of comparing two vectors from different conceptual spaces.

We call analogies like (8.5) and (8.6) *property-based analogies*. Naturally, since the semantic relation between the pairs of terms in these analogies differs from those characterizing analogies that are based on object concepts, explicating them via a search algorithm requires a different approach. In particular, the search space for *X* in an analogy *A:B::C:X* will be the set of lexical items associated with common properties of the concept in *C*. Our proposal for property-based analogies is that their strength depends on two factors: first, on the identification of the dimension(s) that corresponds to the property in the pair and, second, on identifying the degree of *typicality* of that property for the concept in the pair. In other words, we evaluate the aptness of these analogies by checking that the properties in the pairs are from the same dimension and, with the aid of our typicality measure, that they are similarly expected for the concept in the pair. In this sense, an analogy like (8.6) must be considered stronger than the variant FISH:SWIM::BIRD:WALK because, even if birds can walk, flying is more typical than walking for that concept.

From the perspective of a search algorithm, the difficulty lies in identifying the typicality degree of properties for a given concept. In section 2.7.2, we proposed a way of doing this by using distances in conceptual spaces. Let us recall that the conceptual space $C(M)$ of an object concept M includes all the properties that an object falling under M might have.

Now, given an analogy $A{:}B{::}C{:}X$, where A and C are object concepts and B is a property of A in dimension D, the choice for X—which gives the strongest analogical relations—is a different property in D whose typicality degree is closer to B's typicality degree than the typicality degree of any other property in D. We predict that if there are various properties in D with the same typicality degree as B for concept C, then the analogy will be weaker than for concepts for which this is not the case. For example, the analogy LION:BEIGE::RAVEN:BLACK must be judged stronger than LION:BEIGE::DOG:BROWN because several colors other than *brown* are also typical for the concept DOG.

8.6.2 From Properties to Object Concepts

Notice that the search procedure changes if the terms in the analogy appear in a different order, say $A{:}B{::}C{:}X$, where A and C are properties in the same dimension/domain D, and B and X are object concepts. An analogy like RED:APPLE::YELLOW:X requires the agent to search for a concept at the same hierarchical level as APPLE for which *yellow* is highly typical. Instead of typicality degrees, the semantic phenomenon that is relevant here is *diagnosticity* since diagnostic properties allow us to identify concept membership with minimal information.

We claim that in cases like these, the search algorithm involves three steps:

1. Identify the smallest conceptual space M such that $B \subseteq M$.
2. Find a set of concepts in M for which C is typical.
3. Select X from this set as the concept for which C is most diagnostic.

That will be the choice, giving the strongest analogical relation.

8.7 Event-Based Analogies

Next, we focus on a kind of analogy that involves *events* in one way or another. From a cognitive and semantic perspective, the representational structures behind events differ from those behind nouns and adjectives (Gärdenfors 2014, see also Papafragou 2015 for a review). To our knowledge, *event analogies*—as we call them—have only been studied to a limited extent (Goswami and Brown 1990). They come in two types. One concerns analogies of *actions* as in the following examples:

(8.7) WOMAN:WAVE::TREE:SWAY

(8.8) HORSE:GALLOP::MAN:RUN

The second type deals with *results* of events. The following two examples come from the list of stimuli used by Goswami and Brown (1990):

(8.9) BOX:OPEN BOX::BOTTLE:OPEN BOTTLE

(8.10) CHOCOLATE:MELTED CHOCOLATE::SNOWMAN:MELTED SNOWMAN

Our aim is to show that also event-based analogies can be analyzed in terms of conceptual spaces.

We build on the two-vector model of events developed by Gärdenfors and Warglien (2012) (see also Gärdenfors et al. [2018] and Warglien et al. [2012]). The two-vector model states that an event is represented in terms of two components—the *force* (or force pattern) of an action that generates the event and the *result* of its application. Both components are represented as vectors in spaces. In the special case, when there is no change—that is, when the result vector is the zero vector—the event is a *state*.

A central feature of events is that they are based on causal relations: An event contains information about an agent who is the cause of an action that leads to a result related to a patient. The result of an event is modeled as a vector representing the change of properties of the patient before and after the event, as illustrated in figure 8.5.

To model an event, at least two domains are needed: an *action domain* and a *result domain*. The action domain can be conceived as a space of forces or, more generally, force patterns acting upon some patient. The result domain contains dimensions that represent changes in the properties of the patient. The domains represent different types of vectors: Forces have a different nature than changes in properties.

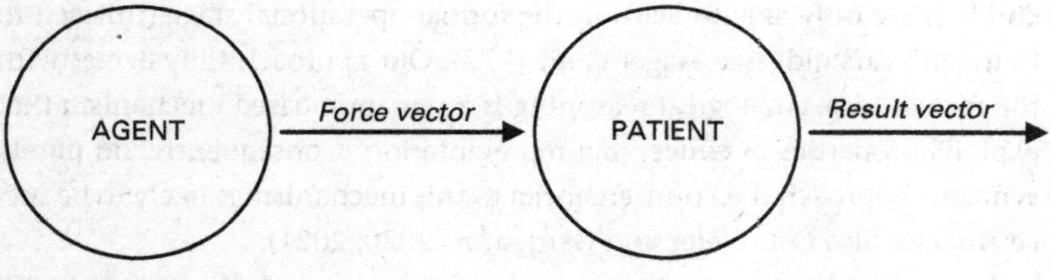

Figure 8.5
The main components of an event representation.

Once the force and results vectors are available, we can use them to model event-based analogies in the same way as in previous sections. For example, in the analogy WOMAN:WAVE::TREE:SWAY, the woman is the agent of an event, and "wave" can be modeled by a force vector that describes the movements of the agent (Gärdenfors and Warglien 2012; Gärdenfors 2014). Then, in the second part of the analogy, the tree is the agent, and "sway" describes a result vector that is the most similar to a woman waving. A similar account can be given for HORSE:GALLOP::MAN:RUN. This analogy works because *run* is the motion verb for *man* whose force pattern representation resembles the most to *gallop* when the agent is *horse*. For similar reasons, we predict that the analogy BIRD:WING::FISH:FIN will be judged as stronger than BIRD:WING::HUMAN:ARM since the force patterns of the movements of wings and fins are, in general, more similar than the force patterns of the movements of wings and human arms. Wings and fins are functional parts in the animal's locomotion and thus exert substantial forces on the surrounding air or water, while arms are typically not used for locomotion.

In the analogy (8.9), "open box" describes a result of an action performed on the box, and the result vector is simply<closed box, open box>. Then, once "bottle" is given, the corresponding result vector is<closed bottle, open bottle>. Thus, the similarity of the result vectors explains the validity of the analogy. A similar analysis can be given for (8.10) and the rest of the analogies studied in Goswami and Brown (1990). Interestingly enough, Goswami (1992) and Goswami and Brown (1990) argue that children's ability to reason with this kind of analogies depends on their semantic competence—that is, on their mastery of the causal concepts involved in the analogy (e.g., *cutting, melting,* and *opening*). As they point out, this contradicts Piaget's idea that analogical reasoning is a formal mechanism that children are only able to learn in the formal-operational stage (thirteen to fourteen years old) (see Piaget et al. 1977). Our approach fully agrees with the former idea. Analogical reasoning is a semantic-based mechanism that exploits properties of conceptual representation. Consequently, no purely syntactic approach (i.e., nonsemantic) to this mechanism is likely to be successful (see also Osta-Vélez and Gärdenfors 2020, 2021).

It is important to note that the role of domains and dimensions is still central in event-based analogies. According to the conceptual space model of events, the meaning of a verb is "a convex region of vectors that depend on a single domain" (or dimension) (Warglien et al. 2012, 172; Gärdenfors 2014). For instance, *paint* denotes a change in the color domain of the patient of the action, and *heat* implies a change in the temperature dimension. Even if the actual actions corresponding to verbs imply changes in multiple dimensions of the patients, verbs work by directing attention to only one of these domains/dimensions. Event-based analogies work in the same way.

8.8 Part-Whole Analogies

As with properties and events, meronomic relations have their own underlying representations, which cannot be fully explained in terms of the above models (Markman 1981; Winston et al. 1987). Analogies that build on these relations exploit properties of the structures of objects and must be modeled accordingly. In what follows, we sketch a proposal for doing this that uses an extension of the conceptual space theory for part-whole relations developed by Fiorini et al. (2014).

8.8.1 Structure Space

The main idea behind this proposal is that meronomic information about an object concept *C* cannot be represented in a standard conceptual space but needs a (higher-level) complementary space called *structure space*. This space encodes two different kinds of information: (1) information about the parts that objects falling under *C* are supposed to have and (2) information about how these parts are arranged in relation to the whole (this is called *configurational* information). The first type is stored in the individual conceptual spaces of each part, while the second is stored in what are called *structure domains*. In this richer conception of a conceptual space, overall conceptual similarity is a function of the similarity measure of the standard conceptual space plus *structural similarity*, which is based on a distance function operating on the structure space.

Our idea is that part-whole analogies depend primarily on structural similarity and, thus, on the information stored in structure space. However, conceptual information can also be used as a reinforcing analogy factor. Consider the following two analogies:

(8.11) HORSE:LEGS::CAR:WHEELS

(8.12) LEG:FOOT::ARM:HAND

The analogy factors behind (8.11) are two: One is configurational and concerns a similarity in the spatial arrangement of the pairs LEGS:HORSE and WHEELS:CAR; the second is conceptual and concerns a *transportation function* that both legs and wheels have in relation to the respective wholes. In (8.12), the only analogy factor is configurational: *Foot* and *hand* occupy the same position in the part-whole structure of *leg* and *arm*, respectively. Besides the great overall similarity in the categories in the analogy, no other salient dimension seems relevant.

8.8.2 Shape Analysis

We will now sketch a possible way of modeling the structural factor in part-whole analogies using ideas from shape analysis. A central aim of shape analysis is to identify structures that remain invariant across the multiple ways in which the shape of an object might appear to an observer. Two things that these structures must capture are the lengths and directions of the parts of an object, which, as is well known in object perception, are often sufficient to categorize it.

Marr and Nishihara's (1978) classic approach to describing biological forms achieves this by representing an object's part hierarchy using cylinder-like modeling primitives. The vector coordinates for these cylinders and their connecting points generate a multidimensional space of shapes, where each particular cylinder shape corresponds to a point. A similarity metric can be defined in this space by taking a weighted sum across all dimensions at different levels. Within this structure space, each point corresponds to a way of combining shapes and relative positions at a given level of decomposition—for example, a specific limb configuration.

Another important component in shape analysis is the connectivity of parts with respect to the whole. Zhu and Yuille (1996) present an approach that emphasizes this aspect, using a model that assumes two fundamental primitive shapes: "worms," which are rectangles with circles at the ends, and "circles" with a deformable angular section that can represent hinges and their flexibility. Like cylinders, a worm can be described by two coordinates: length and width. A hinge between parts can be characterized by its radius and angular values. Various shapes can be represented as compositions of these two primitive structures (see figure 8.6). As in Marr and Nishihara's model, this structure can be represented by a multidimensional vector.

Given a shape represented in this manner, a further structure can be computed by identifying the locus of points equidistant to two or more points in the shape's boundary.[5] This kind of structures are called *topological skeletons* (see figure 8.7); they are invariant representations of the

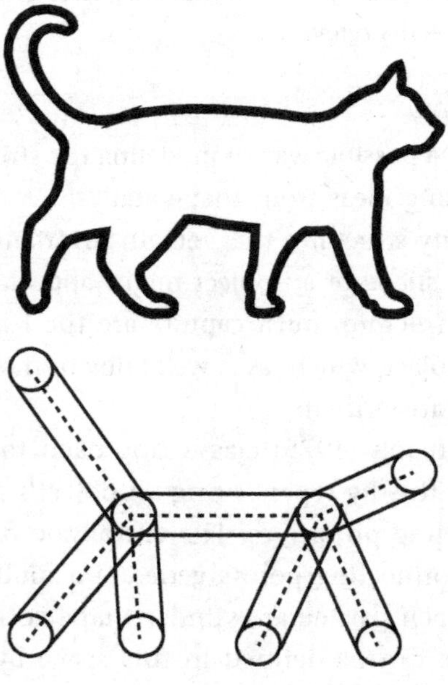

Figure 8.6
Above, the contour and topological skeleton of a cat. Below, its shape composed by Zhu and Yuille's shape primitives.

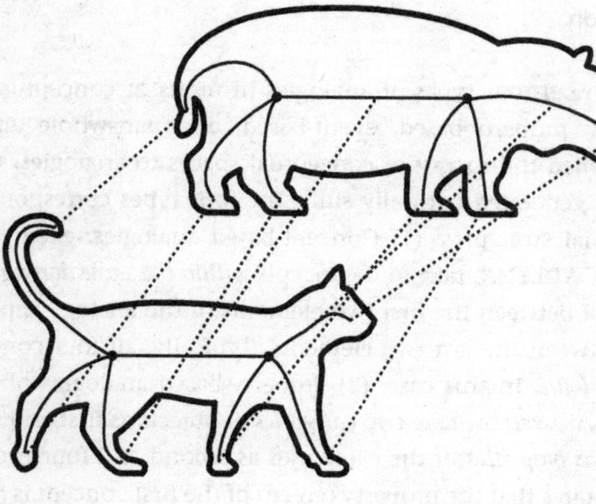

Figure 8.7
Mapping the topological skeleton of a cat to that of a hippo.

connectivity of the parts of an object, and they seem to play an important role in human vision (Lowet et al. 2018; Firestone and Scholl 2014).

Zhu and Yuille (1996, 203–204) present an algorithm that maps the skeletons of different objects onto each other (e.g., see figure 8.7). The algorithm generates an explicit measure of similarity for comparing different shapes or for comparing a shape to a prototypical model. An interesting feature is that their similarity measure works even if the parts of two objects cannot be matched one by one. For example, a hand with a missing finger is still judged to have a high similarity with a typical hand, and a dog with three legs is still seen as a dog.

These outlines of Marr and Nishihara's (1978) and Zhu and Yuille's (1996) models of meronomic shape relations show that, in principle, structural shape relations can be represented in conceptual spaces, albeit high-dimensional and more complicated than for other perceptual properties.

The approach developed by Zhu and Yuille can be used by a search algorithm for part-whole analogies in the following way: Given an analogy *A*:*B*::*C*:*X* such that *B* and *X* are parts of *A* and *C*, respectively, the optimal choice for *X* will be the part of *C* that is mapped onto *B* while matching the topological skeletons of *A* and *C*. In other words, *X* will be the most similar part to *B* in *C*.[6]

8.9 Conclusion

We have analyzed four types of analogies in terms of conceptual spaces: concept-based, property-based, event-based, and part-whole analogies.[7] When we applied the theory of conceptual spaces to analogies, this classification was generated naturally since the four types correspond to different relational structures. (1) Concept-based analogies, exemplified by DOG:WOLF::CAT:LYNX, take four concepts *within the same domain* and say that the vector between the first two elements of the analogy is parallel to the vector between the last two elements. Typically, all four concepts are expressed as *nouns* in this case. (2) Property-based analogies of the type PEAR:GREEN::BANANA:YELLOW take two *categories* of objects as first and third elements and two *properties* of the categories as second and fourth elements. The analogy states that the property (green) of the first concept is *typical* for that domain of the concept and that the property of the second concept is typical for the *same domain*. In this case, the second and the fourth elements are typically expressed by *adjectives*. (3) Event-based analogies, exemplified by HORSE:GALLOP::MAN:RUN, take two categories of objects as first and third elements and two *actions* of the categories as second and fourth elements. The analogy states that the action (gallop) of the first concept involves a similar *force pattern* as the action of the second concept (run). In this case, the second and fourth elements are typically expressed by *verbs*. The results of events can also be compared. (4) Finally, part-whole analogies of the form HAND:ARM::FOOT:LEG exploit information stored in the structure domains of a conceptual space. They depend on a form of structural similarity that compares the parts in the analogies regarding their role in the hierarchy of parts of their respective wholes. We suggested a mapping algorithm that could work as a model of this mechanism, but many others could also work (Siddiqi et al. 1999; Macrini et al. 2011). In this case, the second and the fourth elements are typically expressed by *nouns* that are higher in the meronomic hierarchy than the first and third elements, respectively.

An advantage of using conceptual spaces as a modeling tool is, as we already have noted, that the relative *strengths* of different analogies can be compared. This holds for category-based analogies, for event-based analogies, and for some cases of part-whole analogies. In this way, our models generate predictions that other accounts of analogies cannot make.

An important contribution of our approach is its detailed analysis of the role of semantic similarity in analogy. Propositional-based views, such as Thagard's (Thagard et al. 1990) or Gentner's (1983), also rely on semantic similarity but ignore the modulating role of dimensions and salience. We have shown that focusing on conceptual structure rather than propositional structure has clear advantages for explaining the diversity of analogies and for designing search algorithms.

These search algorithms are amenable to implementations in artificial systems. Since there already exist computational models of conceptual spaces (Chella et al. 2001; Adams and Raubal 2009; Cubek et al. 2015; Lieto 2021), these algorithms could be extended by implementing the search procedures proposed here to account for different types of analogical reasoning. In addition, because conceptual spaces can work as an interface between propositional models and sub-symbolic models of cognition (Gärdenfors 2000), our approach can also be adapted to algorithms based on neural networks (see Jani and Levine 2000). A concrete example is provided by Goldowsky and Sarathy (2024), who introduced computational implementations of concept-based and property-based analogies. For the concept-based analogies, they basically use Rumelhart and Abrahamson's (1973) parallelogram method. A major difference, however, compared to the methods presented in this chapter, is that they do not work with low-dimensional conceptual space but instead use high-dimensional hyperspaces. They present proof-of-concept experimental results for analogies within the color domain.

9 Coherence

A key challenge of our cognitive system is to organize the constant flow of stimuli using limited computational and memory resources. Concepts are crucial to this task, as they allow us to treat different inputs as equivalent, thereby reducing the effort required to process new instances. In this sense, concepts can be seen as cognitive tools that *compress* complex information into manageable, actionable knowledge structures.

Although all concepts simplify information, not all do so with the same efficiency. For instance, the concept AMETHYST can be represented in a compact form because its instances have diagnostic features (*purple hue* and *hexagonal prismatic shape*) that make them immediately distinguishable in almost any context. Its superordinate, QUARTZ, include a much wider variety of instances but remains efficiently compressible: all the members share the same relatively small set of salient dimensions, such as *crystal structure* and *chemical composition*, which are, despite their internal diversity, enough to distinguish them from other minerals. By contrast, categories with less rigid structure and weaker regularities are harder to compress, since greater exemplar diversity requires retaining more information. Consider, for example, the concept GAME (Wittgenstein 2009, §66), whose instances overlap only in few dimensions (e.g., *rules*, *number of participants*, or *scoring system*) but differ across many others, making them difficult to represent compactly. Something similar occurs with very broad concepts, such as LIVING BEING. The diversity of exemplars and wide range of dimensions involved render the concept less compressible and, as we will show later, less useful for reasoning and prediction.[1]

In this chapter, we argue that differences in how easily concepts can be compressed correspond to differences in their *perceived internal unity*—what psychologists refer to as *conceptual coherence*. We claim that one of the aims of our conceptual system is to reduce informational complexity by forming representations that are as compact as possible. The extent to which this compression succeeds determines how coherent a concept is perceived to be when used in language and reasoning.

Several empirical studies have shown that coherence plays a crucial role in conceptual processes, particularly categorization and reasoning. A clear example is the phenomenon of a "preferred level of induction" (already mentioned in section 7.4.2), whereby people tend to favor basic-level concepts over more general ones when reasoning about new instances, since the former are perceived as more coherent than the latter (Corter and Gluck 1992). Similarly, natural concepts (e.g., ANIMAL, PLANT) tend to support stronger inductive inferences because their features are deeply interconnected via causal and structural relationships, while artifact concepts (e.g., TOOL or FURNITURE) have less inductive power since they tend to have weakly interconnected features (Rehder and Hastie 2004; Brandone and Gelman 2009; Hampton et al. 2009). Social categories follow a similar pattern: Patalano and Ross (2007) demonstrated that high-coherence social concepts (e.g., YACHT CLUB MEMBERS) facilitate stronger within-category inferences compared to low-coherence ones (e.g., COUNTY CLERKS), and when reasoning about cross-classified entities, individuals tend to rely on the more coherent category for property transfer (Patalano et al. 2009).

Despite these experimental findings, the theoretical and formal foundations of conceptual coherence remain relatively unexplored. In particular, there is no formal framework that explains why certain concepts exhibit greater coherence than others or how coherence emerges from underlying conceptual structure. This chapter attempts to address this limitation by presenting a novel analysis of conceptual coherence and proposing a method for measuring it. Roughly put, our proposal is that coherence depends fundamentally on the "covariational structure" of concepts—specifically, in the presence of clusters of properties that systematically covary across category exemplars. For example, in the concept BIRD, the consistent relationship between *beak shape* and *feeding habits* across various species illustrates this kind of systematic covariation. The greater the covariance observed within a concept's feature space, the stronger its perceived coherence. Using ideas from principal component analysis, we show that a concept's perceived coherence relates to how easily its structure of features can be reduced to a simpler representation. By offering a formal method to quantify coherence, our approach provides new insights into how coherence constrains inductive reasoning.

9.1 The Nature of Conceptual Coherence

9.1.1 Previous Models of Coherence

Cognitive psychologists have proposed two main explanations for conceptual coherence. The first focuses on *semantic similarity* and claims that concepts are perceived as coherent to the extent that their instances share many features, thus forming a "natural grouping." In other words, greater within-category similarity yields stronger perceived coherence. An example of this view can be found in Rosch and Mervis's (1975) explanation of why basic-level concepts are considered the most "cognitively natural" and are consistently preferred in categorization and induction tasks: Members of basic-level concepts are highly similar to one another and clearly dissimilar to members of the contrast classes, yielding a near-optimal balance between *generalizability*—how easily a property can be projected from a few instances to the entire class—and *informativeness*—the degree to which category membership reduces uncertainty about the properties of a new instance, something that makes conceptual processes more efficient (see section 4.2).

The similarity-based view of coherence faces two important challenges. First, it must provide a criterion determining which dimensions are relevant when estimating pairwise similarity among exemplars. For example, within the concept MAMMAL, DOG may be judged more similar to WOLF than to CAT if the domain *physical appearance* is emphasized, but this will be reversed if a greater weight is assigned to the domain *domestication*. Second, even assuming the relevant domains can be identified, determining their relative importance is challenging since some domains may be diagnostic for certain exemplars but not for others, making it difficult to assign consistent weights across the entire concept (Gelman and Williams 1988). These difficulties have led some psychologists to argue that similarity may be a consequence of conceptual coherence rather than its underlying principle (Medin and Wattenmaker 1987, 28).

The second approach to coherence deals with the limitations of the similarity-based view by grounding concepts in "intuitive theories"

(Murphy and Medin 1985; Gopnik and Meltzoff 1997). An intuitive theory is defined as "a system of interrelated concepts that generate explanations and predictions within a specific domain of experience" (Slaughter and Gopnik 1996, 2967). Such theories determine which domains are relevant and how they should be weighted, thus guiding concept formation, categorization and similarity judgments. Moreover, intuitive theories also account for the role of causality in conceptual structure: In a naive theory of FISH, for instance, *having gills*, *fins*, and a *streamlined body* are seen as causally related features that explain how fish swim. By introducing causal networks into a concept, this approach accounts for why certain features "hang together" in a coherent way.

This view, known as "theory-theory," is not without its challenges. First, the hypothesized intuitive theories seem more complex than the structural properties of the concepts they purport to explain: To possess a theory, a person must first have knowledge of several causal dependencies and explanatory patterns, along with numerous prior conceptual representations, each with its own associated theory. Second, the origins and developmental histories of these theories remain unclear (Sloutsky 2003); the approach replaces the problem of the origin of concepts with the even greater problem of the origin of theories. In addition, multiple theories can account for the same set of observations (Turnbull 2018). Scientists rely on meta-theoretical principles to select among competing theories, but it remains unclear why humans would naturally adopt certain intuitive theories—such as an intuitive biology for classifying organisms—over alternative theories highlighting different causal relationships.

In summary, the two most influential psychological explanations of conceptual coherence face significant limitations—a more recent alternative that focuses on causality will be discussed later in this chapter. As for formal models of conceptual coherence, to our knowledge, only one has been proposed, namely, Thagard's (2002) model. This approach is part of a broader framework designed to accommodate different elements that could participate in coherence relations—not just concepts, but also beliefs, hypotheses, percepts, and propositions.[2]

Thagard's model treats coherence as a *constraint satisfaction problem*. The problem begins with a set E, whose elements can be linked by coherence or incoherence relations. Coherence relations between elements (e_i,e_j) in E are represented as "positive constraints" in a set C^+, while incoherence relations are represented as "negative constraints" in a set C^-. Each constraint is assigned a weight w_{ij} reflecting its relative importance in the problem. The goal is to identify a subset of elements from E that maximizes the satisfaction of constraints—that is, the most coherent arrangement of elements.

To achieve this, E is partitioned into two subsets: a subset A, containing accepted elements, and a subset R containing rejected elements. The weight of the partition $w(A,R)$ is calculated as the sum of the weights of the satisfied constraints. A constraint is considered satisfied if, for any element pair (e_i,e_j) in C^+or C^-, the following holds:

$e_i \in A$ iff $e_j \in A$ or,

$e_i \in R$ iff $e_j \in R$.

The coherence problem is resolved when the partition with the maximum weight is identified.

To illustrate how this model works, consider the coherence evaluation of two categories of water-dwelling animals, each characterized by the following sets of properties (Murphy and Wisniewski 1989):

Category 1: *lives in water, eats fish, has many offspring, and is small.*

Category 2: *lives in water, eats wheat, has a flat end, and is used for stabbing bugs.*

The properties of category 1 align quite naturally: Aquatic animals often eat fish, are small, and produce many offspring; the properties reinforce one another and would form positive constraints with each other. By contrast, category 2 contains conflicting and unrelated properties. For example, *lives in water* conflicts with *eats wheat,* since aquatic animals rarely consume wheat, and *has a flat end* lacks any clear connection to *is used for stabbing bugs.*

When applying Thagard's model, category 1 will yield a higher coherence score because most of its properties can be placed in the "accepted" set A, satisfying numerous positive constraints. Category 2 will struggle to achieve a coherent partition due to its contradictory and disconnected properties, producing more violated constraints and thus lower coherence.

The constraint satisfaction model has been implemented in connectionist networks and applied to a range of domains (e.g., Thagard et al. 2002). It has, however, significant explanatory shortcomings with regard to conceptual coherence. The main limitation is that it does not provide clear criteria for determining when two elements of a concept are compatible or not. In other words, the model does not explain why the features of a concept—the elements in *E*—enter into relationships of coherence or incoherence—a problem that stems from Thagard's avoidance to explicitly define the constraints in sets C^+ or C^-.

Additionally, the model is ambiguous regarding the kinds of entities that may enter into coherence relations. The elements in *E* can represent a single concept and its internal features—as in the approach we will propose—or a set of thematically related concepts, such as DOG-LEASH-VETERINARIAN. This is something that should be clarified, since coherence need not satisfy the same criteria at different scales. In our case, the type of conceptual coherence we analyze concerns the internal structure of concepts.

In the next section, we discuss the relation between the notion of coherence and the notions of *heterogeneity* and *uncertainty*. This will serve as the basis for later introducing our formal model of coherence.

9.1.2 Heterogeneity, Uncertainty, and Coherence

As noted earlier, a concept is coherent to the extent that its diverse instances are perceived as unified. As diversity increases, as is the case with ad hoc concepts such as THINGS TO TAKE TO THE BEACH, this unity becomes harder to grasp. Psychologists call this overall dispersion of variation the *heterogeneity* of a concept (Gelman 2003; Brandone 2017), and it is directly tied to the uncertainty involved when the concept is used. To illustrate, consider what we can infer from categorization alone: Knowing that an object is a *pipe wrench* lets us predict many of its properties with relative confidence, because PIPE WRENCH is a rather homogenous and most instances share a similar shapes, function, and material. In contrast, knowing something is a *handbag* tells us far less, since handbags vary widely in shape, material, size, and purpose, leaving us with greater uncertainty about any new instance. In other words, concepts with lower heterogeneity have greater predictive power and involve less uncertainty when making inferences about new instances (cf. Thagard and Nisbett 1982; Sloman and Lagnado 2005; Brandone 2017).

Conceptual spaces offer a natural way to formalize these ideas. The heterogeneity of a concept corresponds to its volume in a conceptual space—defined as the *size* of the region representing it—that reflects both the variability of properties along each dimension and the number of relevant dimensions. When we categorize an input as falling within this region, the region's volume determines the range of possible property values the input might possess. By gathering additional information or recategorizing it under a narrower (subordinate) concept, we "shrink" the region the input can occupy, thus reducing uncertainty about its unknown properties.

How are heterogeneity and coherence related? Most everyday concepts in natural language involve high levels of property variation, so it is crucial for our cognitive system to find a way to offset this variability and the potential uncertainty it carries. One way of doing this is to look for *structure* in the distribution of properties across the concept, in particular, *covariational structure*. When features within a concept systematically track one another—when knowing some properties reliably predicts others—we can make confident inferences despite the concept's breadth. In other words, covariational structure is what allows us to find unity within diversity, which makes a concept informationally manageable and thus perceivable as coherent despite its heterogeneity.

Natural concepts often exhibit these rich covariational patterns. Within the semantic domain of animals, for instance, a creature's size might covary with specific attributes like *lifespan*, *diet*, or *habitat*. Noticing that larger mammals tend to live longer than insects is a straightforward illustration of this covariation. Similarly, a bird's beak shape can hint at multiple correlated attributes like its *diet*, *habitat*, *nesting behavior*, and even *mating calls*. Observing a bird with a long, slender beak might suggest nectar feeding, which in turn often corresponds to colorful, flower-rich habitats, intricate hanging nests, and melodious calls.

Research by Younger (1990), Billman and Knutson (1996), and Hayes et al. (1996) demonstrates that humans excel at detecting such covariations across various domains. This ability likely has an evolutionary basis: Recognizing patterns in nature was crucial for survival, aiding in the prediction of potential dangers, the identification of food sources, and the choice of viable habitats. As a result, humans appear to have evolved an innate sensitivity to groups of covarying properties. This evolutionary backdrop helps explain why Rosch's (1975) prototype theory emphasizes basic-level concepts as distinctive clusters of strongly covarying features (Holland et al. 1986, 183–184).

This raises the question: Which cognitive mechanisms represent these covariations and how might they work? As mentioned earlier, we speculate that this process might function similarly to principal component analysis. In the next section, we explore how PCA can systematically capture and quantify these covariations, offering a promising way to formalize and measure the internal coherence of concepts.

9.2 Principal Component Analysis as a Measure of Coherence

9.2.1 Basics

Roughly put, PCA is a statistical technique used to reduce the complexity of a dataset while preserving its most important patterns of variation. It does this by defining new axes called "principal components," which capture the largest amounts of variation in the data. The first principal component is the single dimension along which the data varies the most, so the proportion of variation it explains can be seen as an index of how interrelated the features of a concept are.

One way to picture this process is to imagine a cluster of points in a high-dimensional space, each point representing an exemplar with many features. PCA "rotates" this cluster to find a direction where the data vary the most; this direction defines the *first principal component*, and it captures the largest portion of the total variation. Because the first principal component captures the largest portion of overall variation, the proportion of variance it explains indicates how strongly the features covary. In other words, the more variance accounted for by this first principal component, the more coherent (or systematically interrelated) the cluster of points is. Subsequent principal components, orthogonal to the first, capture progressively smaller portions of the remaining variance (see figure 9.1).

PCA is widely used as a dimensionality reduction technique in machine learning, pattern recognition, and data visualization: When datasets have too many variables, PCA helps reduce the number of dimensions while retaining the most important information. The first principal component represents the dimension that provides the most comprehensive "explanation" of the data (see Jolliffe 2002). The relevance of PCA to conceptual coherence lies in how it captures the degree of organization within a set of instances. We propose that a concept's coherence can be estimated by the proportion of total variance explained by the first principal component(s).

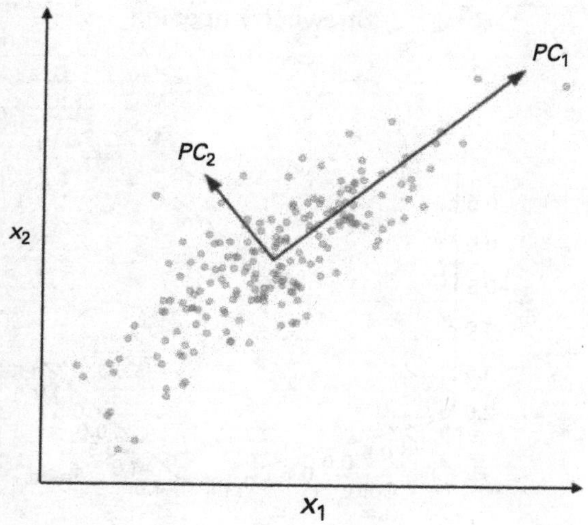

Figure 9.1
The cloud of points represents correlated data. The red arrows show the principal components: PC1 aligns with the direction of greatest variation in the data, while PC2 captures the direction of the remaining variation, orthogonal to PC1.

In a conceptual space, these instances appear as points that may cluster more or less tightly; by isolating the primary axis of variation, PCA provides a way to evaluate how structured the concept is. When the first principal component(s) accounts for a substantial share of the total variance, the concept is perceived as more coherent because its data align along a dominant dimension. Conversely, if the variance is spread across multiple components, the concept exhibits lower coherence, as its instances are more widely dispersed in feature space.

9.2.2 An Illustrative Example

As an illustrative example, let us assume that the concept STRAWBERRY can be represented by only three dimensions: *color*, *taste*, and *size*. Let us now consider a set of observations of strawberries that make up the dataset shown in figure 9.2. The first principal component of the set is the long vector that extends diagonally upward to the right from the mean of the dataset. Orthogonal to that vector, we find a shorter vector representing the second principal component and an even shorter vector representing the third principal component. In this example, the first principal

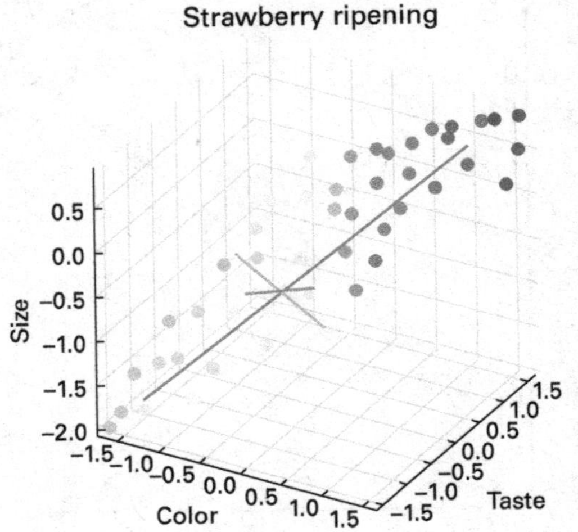

Figure 9.2
An illustration of the first two principal components of a dataset.

component captures considerably more variance than the other two combined and can be used to re-express the data in an efficient manner—without significant loss of information. The dimension identified by the first principal component can be interpreted as a representation of the "ripeness" of the strawberries.

Note that the very fact that we can consistently talk about the "ripeness" of fruits indicates that natural language encodes lexical concepts that refer to multivariate structures as if they were single properties. In essence, the use of this type of multifaceted descriptor promotes communicative and cognitive efficiency; "condensing" diverse information into a single term allows language users to convey and understand complex ideas without needing to specify every detail. This approach aligns with the principle of information economy, which aims to communicate the greatest amount of information with the least amount of effort.

In summary, this example shows how identifying patterns of covariation within a concept increases its perceived coherence. In more complex concepts, different clusters of covarying features might emerge, each contributing its own dimension of coherence; the more variance that can be expressed in fewer principal components, the more compressible the concept will be, and therefore the more cohesive it will feel.[3]

9.2.3 Salience, Covariance, and Correlations

Unlike the simplified example above, most real-world concepts span numerous dimensions that differ in importance or salience. For example, in the concept DOG, dimensions like *size* or *coat* are typically more influential than features such as *body temperature*. In other words, certain dimensions, though present, do not meaningfully impact how we classify or understand the concept. This raises a challenge for the conventional PCA procedure, which assumes that every dimension matters equally. To address this, we now examine how PCA works in detail and show why incorporating salience weighting is necessary for capturing a concept's internal organization.

The standard method for PCA proceeds as follows (Jolliffe 2002): First, the range of (continuous) variables is standardized to ensure that no single dimension dominates the analysis due to differences in scale. Next, a covariance matrix is computed to capture how pairs of variables vary together, highlighting potential correlations among dimensions. Finally, the eigenvectors and eigenvalues of this matrix are calculated to determine the principal components—where eigenvectors represent the *directions* of maximal variance and eigenvalues indicate the *magnitude* of this variance. The principal components are ranked by their corresponding eigenvalues, with the first principal component accounting for the greatest variance in the dataset.

The first step of standard PCA assumes that all dimensions are equally important, since it standardizes each variable's scale before computing correlations. However, in psychological representations of concepts (as well as in conceptual spaces), dimensions differ in salience: Some properties contribute more to a concept's structure than others. This suggests that principal component calculations should incorporate salience values rather than treating all dimensions as equivalent. Notably, this also implies that variable ranges should not be standardized, since standardization artificially equalizes their influence. To address this, we propose weighting covariations according to salience rather than relying on correlations derived from standardized variable ranges. This ensures that the principal components better reflect the structure of the concept rather than merely reflecting statistical variance.

It is important to note that covariance measures require a reference point because they are based on deviations from that point. Standard PCA implicitly uses the mean as this reference when standardizing data.

However, since we reject standardization precisely because dimensions differ in psychological importance, we must explicitly choose a cognitively meaningful reference. To do this, we introduce the idea of a *universal prototype* (representing the prototype of "thing"), which we represent by the *mean* of the positions of all objects. This universal prototype serves as a more nuanced reference for understanding how different features relate within a concept. We take the universal prototype to serve as the origin of the conceptual space. The role of the universal prototype becomes particularly clear when considering concepts that exhibit little or no variation in most of their dimensions. Even when individual dimensions do not correlate strongly, covariance remains meaningful because it is assessed relative to the entire conceptual space. Many animal concepts, for example TIGER, illustrate this point. For such categories, there will be little or no direct correlation between their dimensions, yet their overall covariance structure remains stable when measured against the universal prototype. This is the reason why we use *covariance* rather than *correlation* to determine the coherence of a concept.

With this framework in place, we can now introduce a concrete measure of conceptual coherence. Given a set of exemplars of a concept X and its salience-weighted covariance matrix Σ_w, we apply PCA to Σ_w to obtain eigenvalues $\lambda_1, \lambda_2, \ldots, \lambda_d$ ordered from largest to smallest. These eigenvalues quantify the variance explained by each principal component. Let

$$p_i = \frac{\lambda_i}{\sum_{j=1}^{d} \lambda_j}$$

be the proportion of the total variance explained by the ith principal component. We then define the *coherence score* of X as

$$C_X = \sum_{i=1}^{d} w_i p_i$$

where the w_i are positive weights that decrease geometrically with i (e.g., $w_i = (1-r)r^{i-1}$ for some $0 < r < 1$) and are normalized such that $\sum_{i=1}^{d} w_i = 1$.

This particular weighting ensures that earlier principal components (those capturing the most psychologically relevant variance) contribute proportionally more to the score, while later components have a progressively smaller influence. For highly coherent concepts, in which most variance is concentrated along the first few components, C_X will be high; but for less coherent concepts, whose variance is distributed more evenly across many components, the weighted sum will result in a lower score.

9.2.4 Some Properties of This Coherence Measure

We can now examine the five key features of our coherence measure. First, the measure is dependent on the salience values of the domains. If a domain has low salience, then all volume related to it will be small, and, consequently, its contribution to the total coherence of the concept will be small. For example, consider the reclassification of the concept WHALE from FISH to MAMMAL. In the earlier classification, the "ecological" domain (*lives in water*) was highly salient, overshadowing key biological properties such as *warm-bloodedness*, *live birth*, and *milk production* (Gärdenfors 2000, section 6.4). However, by shifting salience toward these biological domains and away from purely ecological ones, WHALE became better aligned with the core properties of MAMMAL. As a result, its internal organization became more coherent.

Second, our measure does not rely on the probabilities of individual properties; instead, it uses the *geometry* of how property dimensions overlap in the conceptual space. This focus on relational structure—how tightly clustered or well aligned the instances of a concept are—does not require detailed probabilistic data. As a result, the measure can be applied in contexts where exact probability distributions are unavailable or difficult to estimate.

Third, concepts spanning multiple salient domains generally attain higher coherence than those with fewer salient domains. This helps explain why natural concepts, such as BIRD, often exhibit strong coherence: They involve multiple correlated dimensions—like *size*, *beak shape*, *habitat*, and *diet*—that systematically covary. In contrast, artifacts such as CHAIR or CLOCK typically have fewer salient attributes, with physical features varying widely and only function remaining highly relevant. Consequently, their dimensions exhibit weaker covariance, resulting in lower overall coherence.

Fourth, even among concepts with multiple salient dimensions, coherence increases further when these dimensions form tightly interrelated clusters of properties. Such clusters are often present in natural kinds, making them highly coherent and more easily learnable (Billman and Knutson 1996; Kornblith 1993; Shipley 1993). By contrast, artifacts often lack such tight clustering, which explains their typically lower coherence and higher variability.

Fifth, it follows from our definition that superordinate concepts are less coherent than their subordinates: Even though their broader scope can exhibit higher pairwise covariation, the large volume disperses variance across many principal components, so the first few capture a smaller proportion of the total.

We have shown that PCA, when modified to incorporate dimensional salience and a universal reference point, offers a method for estimating conceptual coherence. This framework helps explain key cognitive phenomena, such as why natural categories tend to be more coherent than artifacts and why concepts with tightly clustered properties are easier to learn. In the following section, we explore how coherence relates to causal knowledge and its role in inductive reasoning.

9.3 Coherence, Causality, and Induction

9.3.1 Coherence and Causality

Some researchers argue that coherence cannot be fully explained by covariational learning alone and suggest that causal knowledge plays a central role instead. In a series of experiments, Malt and Smith (1984) and Ahn et al. (2002) found that typicality judgments—assessments of how representative an exemplar is of a given concept—were stronger when the exemplar contained pairs of causally related properties. Focusing on the concept BIRD, they presented subjects descriptive exemplars that shared general features like *having wings* but differed in additional properties: Some had *lives near the ocean* and *eats fish*, while others had *is white* and *eats fish*. Although both property pairs exhibited similar covariation, participants rated exemplars with *lives near the ocean* and *eats fish* as more typical, because they perceived the properties as causally linked. The researchers concluded that semantic judgments depend not only on pairwise covariation but also on causal relationships, a finding consistent with ideas the theory-theory, particularly its interpretation of conceptual coherence.

There is, however, an alternative explanation to these experiments (see also Rogers and McClelland 2004). Participants might give more weight to features they perceive as causally linked because they coherently covary with multiple other properties. For instance, *lives near the ocean* and *eats fish* apply to many shorebirds sharing other common traits like *dives* and *can swim*. Conversely, *is white* and *eats fish* could apply to various birds—for example, herons—making this combination less coherent. The upshot is that causality might not be essential to explain coherence. Instead, the perceived coherence among two properties may stem from their covariation with other properties, contributing to the principal components and thereby enhancing coherence. This aligns with the measure we propose.

Moreover, our perception of the connections between property pairs within a concept may be influenced not only by their covariation but also by the relationships among properties across similar concepts within the same conceptual domain. For instance, subjects might perceive the pair *is white* and *eats fish* as less correlated because they recognize that *color* and *feeding habits* are not consistently related across different animal concepts. In contrast, the pair *lives near the ocean* and *eats fish* may be viewed as more correlated, as habitat and feeding habits often exhibit a strong relationship across various animal concepts. This broader perspective suggests that concept coherence is shaped by the statistical structure of larger groups of concepts, not just within a single concept (cf. Billman and Dávila 2001).[4]

Following a similar line of argument that emphasizes causality as the key factor behind coherence, Rehder (2017) and Rehder and Kim (2006, 2010) examined the "coherence effect" in categorization tasks. They found that people are more likely to judge an exemplar of a concept as a good category member if it preserves the *causal laws* of that concept, that is, the causal relationships between features that participants had previously learned.

As an example, consider the concept TROPICAL FROG: Suppose people have learned that being poisonous (feature A) causes brightly colored skin (feature B) in these frogs. Given this causal relationship, people judge exemplars of frogs that are not poisonous and not brightly colored (¬A¬B) as better category members than exemplars that are either not poisonous but brightly colored or poisonous but not brightly colored (¬AB or A¬B). This preference arises because ¬A¬B exemplars preserve the expected causal structure ("no poison, therefore no bright color"), whereas ¬AB and A¬B violate it, as their features do not align with the learned causal relationship.

Our approach can account for this effect by considering the role of *loadings* and *scores* in the relevant PCA-reduced conceptual space. In PCA, each principal component is defined by a set of loadings—that is, weights taken from the eigenvectors of the covariance matrix—that quantify how strongly each original dimension contributes to that axis. When two dimensions share large, same-signed loadings, they co-vary, and the component points along their joint direction. The *scores* are the coordinates of the exemplars when projected onto the new axes; they show how well an exemplar aligns with that covariation pattern. When subjects learn the causal law "Poison causes bright color," the dimensions encoding these two properties become strongly positively correlated in a (hypothetical) sample of FROG that learners generate from the rule. PCA therefore places this relation on the first component, with both dimensions receiving large, same-signed loadings. Exemplars that are both poisonous and bright color (AB) are projected far along the positive side of the first PC, while those that are neither (¬A¬B) are projected far along the negative side. Mismatched cases (A¬B or ¬AB) cancel out on this component and instead receive large scores on an orthogonal "mismatch" axis.

On this account, coherence effects reflect how closely exemplars align with the dominant component: Rule-consistent cases have strong scores and appear more coherent, while mismatched cases lie closer to the origin, with weaker alignment, and are judged less typical.

To be clear, we do not claim that causality is irrelevant to conceptual coherence. However, we believe that outside of expert knowledge, most causal understanding of inter-feature relations is likely "shallow." For instance, while most people acknowledge that the curved shape of a boomerang or the size of a bird's wings play a causal role in—or are *enabling conditions of*—their respective flight patterns, comparatively few individuals really grasp the underlying mechanisms that govern these causal relationships.

We believe that much everyday causal knowledge is acquired through *causal generics*, that is, statements like "Birds can fly because they have wings." Although these kinds of statements only highlight causal relationships without accounting for the underlying mechanisms, they provide two types of valuable information: (1) they signal a positive pattern of co-occurrence between two dimensions of a concept (*wings* and *flight* tend to co-occur), and (2) they establish an expectation of asymmetric intervention (removing wings eliminates flight, but inhibiting flight does not affect the presence of wings). In this way, they act as cognitive shortcuts that allow subjects to internalize both the correlation and its causal direction without needing to infer them from experience through repeated examples. We speculate that many folk concepts are not formed by a purely data-driven process but are structured around causal generics that establish the covariational structure, which later learning tests and refines. Still, coherence is more dependent on detecting covariation between properties than on an explicit understanding of causal relationships.

In summary, while causal knowledge plays a role in how we perceive conceptual coherence, its influence is primarily confined to expert knowledge. In everyday reasoning—particularly for *folk concepts*—coherence is guided more by statistical mechanisms that identify patterns of covariation and plausible enabling conditions rather than by an explicit understanding of causal relationships.

9.3.2 Coherence and Induction

Empirical studies have shown a positive correlation between coherence and the confidence with which concepts are utilized in inductive tasks. In an influential study, Gelman (1988) demonstrated that by second grade, children make more inductive inferences with natural concepts (e.g., RABBIT, APPLE) than with artifact concepts (e.g., CHAIR, CLOCK), especially at the basic level, whereas preschoolers show no difference. She explained this shift in terms of the emergence of domain-specific knowledge that modulates simple similarity judgments. On our view, learning domain-specific knowledge is learning covariance structure. Because many biological kinds exhibit richer, more integrated covariances than common artifact categories, they support broader and stronger within-category projections (see section 5.4.3).

Rehder and Hastie (2004) further noted that inductive generalizations tend to be stronger when based on coherent concepts. They proposed that the *typicality effect* observed in category-based induction can be interpreted as a coherence effect: Atypical exemplars of a concept support weaker generalizations than more typical ones because they violate expected correlations associated with that concept. For instance, although penguins possess feathers and wings, their inability to fly makes them atypical birds, which weakens the inductive inferences drawn from them.

Coherence also plays a role in *between-category induction*. As we explained in chapter 7, in category-based induction people rely heavily on overall similarity between two concepts to project a property from one to the other (Douven et al. 2023). However, when the projected property is known rather than blank, property effects might modulate similarity judgments.

Heit and Rubinstein (1994) showed that when making inferences about animal behavior, such as *prefers to feed at night*, participants found TIGER → HAWK inferences more compelling than CHICKEN → HAWK inferences, even though overall similarity between the concepts indicates otherwise (see section 7.3.5). This suggests that people check whether the concepts in the premise and in the conclusion share a cluster of covarying features that would make the property plausible. The effect depends on coherence because, without stable clusters of covarying features, there would be nothing meaningful to reweight toward.

In the next section, we explore how coherence influences reasoning in the domain of social concepts, particularly in relation to the conjunction fallacy.

9.4 Coherence, Social Concepts, and the Conjunction Fallacy

So far, we have sought to explain why our cognitive system strives to construct and prioritize coherent concepts, using natural kind concepts as the primary example. However, we also argue that social concepts—such as BANKER, GUITARIST, CONSERVATIVE, and FEMINIST—are especially influenced by coherence and, consequently, by correlations (Patalano et al. 2006). Experimental evidence supports this claim. For instance, Nguyen and Chevalier (2015) discovered that five-year-olds favor coherent concepts when making inductive decisions about social concepts. In one of their experiments, participants were told the following: "Baseball players like apples, and board game players like bananas. This is Pat. Pat is both a baseball player (coherent concept) and a board game player (incoherent concept)." When asked whether they believed Pat would prefer an apple or a banana, most chose the fruit linked to the coherent concept.

The use of social and natural concepts in categorization relies on different principles and provides different kinds of information. Natural concepts are typically defined by observable features, both behavioral and physical, and remain relatively stable across various cultural and social contexts. In contrast, social concepts are defined by socially constructed characteristics or roles, they vary regarding the types of features they denote—from patterns of observable behavior to doxastic and ideological dispositions—and they show less stability over time and across cultural contexts compared to natural concepts.

More importantly, the two kinds of concepts have distinct informational profiles. Natural concepts typically give rise to rather precise expectations about the distribution of properties among their exemplars and resists cross-classification at the same hierarchical level. Social concepts, by contrast, allow for a high degree of cross-classification: Individuals often belong to multiple social categories simultaneously. Even when stereotypes are used to manage the variability encountered in social categorization, no single social concept exhaustively determines the other traits a person may have. For instance, someone categorized as a SCIENTIST can be also categorized as a MUSICIAN, a PARENT, a LIBERAL, or an INTROVERT as social roles and personal traits overlap freely.

This suggests that when we learn and reason about people, our process resembles "inference learning" (Jones and Ross 2011; Yamauchi et al. 2002).[5] That is, we attempt to predict which additional social categories an individual might belong to by drawing on our general knowledge of correlations and selecting those that best align with our prior understanding of the person. In essence, if we have certain prior knowledge about an individual x, we expect x to exhibit properties that are positively correlated with this knowledge and to lack properties that are negatively correlated with it.

This idea offers a straightforward explanation of the conjunction fallacy from a *coherentist* perspective. Repeating our presentation in section 5.6.3, Tversky and Kahneman (1983) presented participants with a description of Linda, a thirty-one-year-old woman who was single, outspoken, highly intelligent, and a philosophy major. As a student, she was deeply engaged in issues of discrimination and social justice and participated in anti-nuclear demonstrations. Participants were then asked to assess the likelihood of Linda being a bank teller versus Linda being both a bank teller and a feminist. In contrast to the laws of probability, most participants found that Linda being a feminist bank teller was the most likely scenario, demonstrating a "conjunction fallacy."

From our perspective, this fallacy arises because the attribute of being a feminist strongly correlates with the characteristics previously described about Linda. The scenario in which Linda is a feminist bank teller enhances the conceptual coherence of our representation of Linda more than if she were merely a bank teller. In other words, if our cognitive system prioritizes conceptual coherence, it becomes more natural to classify Linda as a feminist bank teller, as this aligns better with the provided information. Conceptual coherence thus generates the expectations that underlie our explanation of the Linda problem in section 5.6.3. Consequently, our two accounts reflect complementary aspects of a single explanatory framework.

By contrast, categorizing her as only a bank teller introduces uncertainty about additional properties she might have—particularly those positively correlated with being a bank teller (e.g., *not being a feminist*) but negatively correlated with our prior knowledge of Linda. In other words, people, driven by an inclination for coherence, rely on clusters of covarying attributes when forming conceptual representations. As a result, they may override statistical reasoning to construct a more internally consistent image of Linda based on the given description.

A similar approach to the one we are proposing was developed by Siebel (2002), who builds on Thagard's coherence model (cf. Trpin and Hartmann 2024). An interesting aspect of Siebel's analysis is that he links coherence (from the perspective of classical epistemology[6]) to inference and then to explanation. According to Siebel, people judge Linda as a feminist bank teller because this classification establishes more "inferential connections" with the prior information about her, thereby enhancing our ability to generate explanations of her characteristics.

From our perspective, the connection to explanation arises naturally, as inferential connections ultimately reduce to—and can be understood in terms of—covariation patterns embedded in conceptual structures. Rather than relying on a separate inferential mechanism, we view these connections as a consequence of the way concepts encode correlations between properties.

Interestingly enough, experimental evidence on the cross-classification of people and induction supports the previous ideas (Patalano et al. 2003; Patalano et al. 2006). People exhibit a strong preference for using high-coherence concepts as a basis for explaining people's features, even when doing so leads them to violate the laws of probability.

9.5 Conclusion

In this chapter, we have proposed a novel analysis of conceptual coherence and a method for quantifying it. We claimed that coherence is related to the compressibility of the covariational structure of concepts—that is, from the way clusters of properties systematically co-occur across category exemplars and can be efficiently compressed into a small number of principal components.

This framework contributes to research on concepts in several ways. First, it clarifies the relationship between coherence and inductive reasoning by explaining how more coherent concepts support stronger inferences. Second, it reconciles competing accounts of conceptual structure by showing that both similarity and causal knowledge may influence conceptual coherence. Third, it offers a computationally tractable model that could find applications in cognitive science, artificial intelligence, and semantic modeling.

Our proposal also accounts for different empirical findings in categorization and reasoning. It explains why natural kind categories exhibit stronger coherence than artifacts, why concepts with tightly interrelated features are easier to learn, and why social concepts influence probabilistic reasoning, particularly in cases such as the conjunction fallacy. Our approach provides a principled explanation for these phenomena by linking coherence to underlying statistical structures rather than treating it as a primitive notion.

Something that needs to be noted regarding our model is that while PCA can serve as an effective tool for quantifying coherence, alternative methods—such as mixed-variable factor analysis or multiple correspondence analysis—may be better suited for concepts involving both continuous and discrete features. Future work could explore these approaches to refine our understanding of coherence across diverse conceptual domains, ensuring that the proposed framework captures the full complexity of conceptual organization.

In the following chapter, we will deepen our analysis by exploring *concept essentialism*—the idea that certain concepts possess an inherent nature. By examining how the covariational structure of concepts shapes our perceptions of such essences, we aim to uncover additional factors that contribute to conceptual coherence.

10 Essences

> On ne voit bien qu'avec le coeur. L'essentiel est invisible pour les yeux.
> —Antoine de Saint-Exupéry, *Le petit prince*.

10.1 Background

The idea that things have essences goes far back in the history of philosophy. Aristotle distinguished between accidental properties, which objects possess contingently, and essential properties, which define an object's fundamental nature, identity, and powers. This distinction allowed for definitions. Centuries later, Locke (1689/1975) made an alternative proposal, differentiating between "nominal" essences—our mental constructs based on observable features—and "real" essences, representing an object's underlying physical constitution, often beyond our sensory perception. Locke described real essences as "the very being of anything, whereby it is what it is . . . the real internal, but generally . . . unknown constitution of things, whereon their discoverable qualities depend" (III.iii.15).

Contemporary cognitive psychology has taken up the debate on essentialism, focusing on the question of why humans conceptualize objects in terms of essences. Gelman, a leading figure in this discussion, argues that (cognitive) essentialism is a "universal habit of mind" (2003, 15) which comprises three main components (2003, 7):

1. A belief that certain categories are natural kinds.
2. A belief that some unobservable property *causes* the observable similarities shared by category members.
3. A belief that everyday words reflect the real structure of categories.

There is strong evidence that essentialist thinking develops in children as an early cognitive bias. Gelman (2003, 297) writes that "essentialism falls out as a consequence of children being very good at a number of other skills: distinguishing appearance from reality, searching for causes, noting correlational clusters, and understanding how labeling works." Following Gelman, we focus on studies of children since they best reveal to what extent essentialism is "a habit of mind." Their essentialist thinking reveals a powerful capacity to look beyond perceptual features when forming object concepts.

Cognitive scientists working on concepts often put special focus on the role of perceptual features in conceptual processes (e.g., Gärdenfors 2000). Essentialism presents a challenge for such work, as the properties considered essential to a concept frequently lack direct perceptual correlates. For example, dogs vary a lot in their looks; nevertheless, already small children believe that there are some "inner" feature that constitute the essence of dogs (Newman et al. 2008).

The challenge of essentialism lies in identifying these inner features. In other words, the task is to specify how our beliefs about "essences placeholders"[1]—the fundamental quality that defines membership in a category associated with a concept—are shaped. Another problem is that the ascription of essences varies across different types of concepts.[2] For instance, animals are typically seen as having more essential properties than artifacts (for a review, see Neufeld [2022]).

This can be explained using the same approach we used to explain the notion of coherence in the previous chapter. Roughly, animal categories have many properties that go together, whereas the same is not true for most artifacts. This idea forms the basis for our subsequent analysis.

In this chapter, we propose that the essentialist bias stems from a principle of cognitive economy that aims to lessen the complexity of our concepts. This is accomplished by identifying and compressing clusters of correlated features into singular, multifaceted new variables. Our main thesis puts the principal components of a concept back in focus: Apart from being markers of the degree of coherence of concepts, principal components can be interpreted as "hidden variables" that play the role of the "essence placeholders" of concepts.

We use the theory of conceptual spaces as a framework to express these ideas and demonstrate that principal components can explain several aspects of empirical investigations concerning essences. In particular, we focus on causal cognition. We also discuss how to interpret functional properties in artifact concepts and examine the connection between generic language and essentialism from a developmental perspective. The chapter concludes by discussing the relationships between generics and essentialism, addressing why generics emerge early in linguistic development.

10.2 Covariation, Again

As noted in the previous chapter, our ability to detect patterns is central to how we understand and conceptualize the world. We have evolved an especially acute capacity to register covariations among features due to their utility in helping us make sense and predict our environment. Indeed, psychologists have shown that identifying covariations among features across domains is key to concept formation (Medin 1983; Younger 1990; Billman and Knutson 1996). Yet, as Kornblith argues (1993, 105–106) only those clusters of sensory properties that are organized in a stable enough way cohere into natural categories capable of supporting inductive inference. Animal concepts supply the clearest example: Within them, properties tend to occur and covary substantially, creating a network of interconnected features.

Indeed, Shipley (1993) compellingly demonstrated that children's knowledge of animals exhibits a higher degree of organization compared to other semantic domains. This organization stems largely from the *consistent contrast* between property clusters across different animal concepts. For instance, wolves (meat-eating, howling, thick-furred, forest-dwelling) contrast sharply with parrots (seed-eating, squawking, feathered, rainforest-dwelling) and frogs (insect-eating, croaking, moist-skinned, wetland-dwelling). These systematic distinctions facilitate the identification of covarying property clusters among similar categories and improve the overall organization of the domain—something that, as Billman and Dávila (2001) experimentally demonstrated, speeds up the leaning of new similar categoris. Furthermore, the identification of clusters of covarying properties is crucial for inductive reasoning. As Gelman (2003, 316) articulates, "The more commonalities you have learned about a category in the past, the stronger your inferences about that category are in the future. Metaphorically, property clusters are magnets: the more mass they accumulate, the more strongly they pull in new properties. Inductive success is rewarded with more induction."

Empirical evidence consistently indicates that animal concepts are regarded as possessing rich essences (Gelman 2004). We believe that, as with coherence, this is due to their rich structures of covariant properties. Artifacts concepts serve as a useful contrast case here. Many features in artifacts occur *superfluously*; they tend to vary independently and are often unrelated to the object's core function—for instance, the color of a clock bears no intrinsic relation to its function, shape, material, accuracy, or mechanical properties. As a result, and as empirical studies indicate, artifacts are perceived as having weaker or more limited essences, primarily defined by their intended function rather than by a tightly connected set of features.

In brief, concepts with covarying properties are more strongly expected to have fundamental natures or essences. In the next section, we will explore this idea in detail.

10.3 Our Proposal

Formal models of concepts in the literature struggle to capture the ideas discussed earlier. As a result, the issue of essentialism has been largely overlooked in this field of study.[3] We propose that conceptual spaces are well suited to address this gap, particularly in elucidating the problem of locating the "essence placeholder" within concepts.

Remember that in a conceptual space, the exemplars of a concept are represented as points within a convex set. The distribution of these points may vary. When a significant number of points exhibit similar "values" for a particular set of dimensions/domains of the conceptual space, they form a cluster in a specific subregion, expressing a covariation among these domains. A concept can encompass numerous dimensions/domains, potentially resulting in substantial computational costs. A mechanism or heuristic for reducing conceptual complexity is desirable for informational economy and tractability.

As proposed in the previous chapter, we suggest that this mechanism functions analogously to PCA: It reduces the concept's dimensionality by identifying clusters of covarying dimensions and compressing them into new variables that maximally capture variability.

Paralleling our analysis of coherence, the essence of a concept can be characterized by its first principal component (or components). Specifically, the degree to which the first principal component reduces the variance in the dataset associated with a concept correlates with the strength of its essence. By definition, greater covariance among the dimensions of the dataset results in a larger reduction of variance by the principal component. Our hypothesis suggests that a principal component yielding a substantial variance reduction generates a strong "essential" variable, while one producing less reduction creates a weaker essence. Given that we also employ principal components to assess coherence, it follows that a strong covariation exists between a concept's coherence and its essence.

Our approach offers a technical interpretation of a notoriously ambiguous concept. Central to the definition of essentialism as found in the literature is the proposition that the essence of a concept serves as a "unifying principle" that explains the observable features of its exemplars (Newman and Knobe 2019, 2). Nonetheless, describing this principle without invoking enigmatic entities poses a significant challenge. Our solution is to operationalize this principle in terms of covariation. Roughly put, the development of essentialist beliefs about categories hinges on the capacity of our cognitive system to identify clusters of covariating domains[4] that can then be 'condensed' into a single variable (or a few variables) that retains comparable or identical explanatory leverage regarding the structure of a dataset.

Consider the concept PERSONALITY as an illustration of this essentialist approach. What we call a person's personality—often seen as her "essence"—is treated as an internal constant that compresses a wide range of observable behaviors into a single, coherent, variable. Although this variable can be decomposed into narrower components like the "Big Five" personality traits (Briggs and Cheek 1986), it continues to function as a unified summary of an individual's behavioral dispositions.

It should not be assumed that all concepts have an essence. When the properties of a concept are unrelated and more or less equally weighted, the principal component will be very small, and therefore, the concept will not have any identifiable essence. In the STRAWBERRY example in the previous chapter, only three domains were involved; but most natural concepts draw on far more, many of them only marginally relevant or not relevant at all—for example, even though all objects have a degree of radioactive radiation, this is typically so small that this domain does not influence the categorization of an object.

In the STRAWBERRY example, it was possible to interpret the principal component as representing a dimension of ripeness. But when there are too many variables at play, the leading components become harder to interpret, and there may also be more than one component involved in what is judged to be the essence of a concept. We do not claim that *ripeness* is part of the essence of a STRAWBERRY (although an empirical investigation could show that it is); we only used the example to illustrate the notion of a principal component. For most natural concepts, the principal components are difficult to identify explicitly—let alone to give names, yet they exist for any given dataset and, on our view, offer a way to make precise what we intuitively take to be a concept's essence.

10.4 What the Principal Components Explain

This section examines evidence supporting the hypothesis that both children and adults mentally represent essences, a view that accounts for several inferential patterns observed in children (Keil 1989; Gelman 2003) and adults (Rips 1989). These inferences fall into two types: categorization and induction. As the properties of a concept cluster more closely, the first principal component explains more variance, allowing for more inductive inferences. Thus, categories with rich essences have greater inductive potential.

10.4.1 The Essence of Natural Kind Concepts

As noted earlier, essentialism is particularly strong in children's thinking about animals (Gelman 2003, 49; Keil 1989). The fact that animals are living things shows up in many common properties: They move, eat, breathe, have offspring, are born, die, and so on. Children seem to assume that there are some underlying dimensions for all these properties, even if these dimensions are neither observable nor possible to express explicitly (Kornblith 1993). This is a characteristic of principal components: While the perceptible properties of an object may change over time, the values for the principal components are typically less varying. For example, an insect can change from a larva, via a cocoon, to a butterfly. These stages of the butterfly are perceptually radically different, while the essential variables relating to living organisms are constant.

Since principal components maximize explained variance, a small number of them typically account for more of the observed similarity among a the instances of a concept than any single observable feature. The leading components are determined by the covariance structure of the features, not by any one feature, and they usually do not align with directly observable dimensions. This is why the essence of a concept is often thought of as being something hidden "inside" the objects falling under the concept.

A further reason to explain why essentialism is stronger for animal categories and weaker for artifacts is that the strong coherence of the properties related to "living" in animals has no correspondence for artifacts. This entails that animal concepts allow inferences based on hidden properties, but artifacts do not. A consequence is that transformations that radically change the appearance of an object result in judgments of concept change for artifacts, but animal categorization remains stable.

An example of this effect comes from Keil's (1989) investigation of transformations of objects that change their appearance. For animal concepts, such changes do not affect children's conceptualizations. For example, a skunk that looks like a raccoon but has the same inside as a skunk is still judged to be a skunk. In contrast, children judge changes in the appearance of artifacts as more important for how the artifact is conceptualized. For example, a coffee pot made of a birdfeeder is more likely to be classified as a coffee pot than a birdfeeder (Keil 1989, 176–181).

The essence of an animal is difficult to make explicit, but it seems to relate to the factors that are necessary for the animal to be living and having agency, being able to act on its own independently of external forces. In contrast, plants are seen as living and growing but lacking agency.[5]

10.4.2 Artifacts Have Weaker Essences

Most artifacts, such as chairs, clocks, and telephones, are categorized based on their functional properties (Kelmer Nelson 1999; Mandler 2004)—that is, what we *do* with them. Actions also seem central to the development of concepts in infants and children. Nelson (1996, 110–111) summarizes the position:

> The claim was that infants would form concepts of objects that were functional in their own lives and would, therefore, learn to name them easily and early. The object's function from the child's point of view (e.g., rolling a ball) would form

> the core of the concept, and its perceptual appearance would be added as a cue to the identification of instances.

For example, when deciding whether an object is a chair, the perceptual dimensions of the object, like those of *color*, *texture*, and *weight*, are largely irrelevant or at least extremely variable.

We propose to analyze functional properties by reducing them to the actions that the objects "afford" (Gibson 1979). To continue with the example, a chair is prototypically an object that affords back-supported sitting for one person—that is, an object that contains a stable flat surface at a reasonable height from the ground and another flat surface that supports the back (see also Jackendoff 1987, 105). In contrast, a sofa affords seating for more than one person, support to the back, and lying down, but affords less to stand on. A stool affords seating for one person, offers no support for the back, but better affordance to stand on.

In support of this analysis, Vaina (1983, 28) writes, "The requirement for efficient use of objects in actions induces strong constraints on the form of representation. Each object must first be categorized in several ways, governed ultimately by the range of actions in which it can become involved." Following this idea, we propose that function concepts be interpreted with the aid of the action domain. The following special case of the thesis about properties was presented by Gärdenfors (2007):

> *Thesis about functional properties*: A functional property of an object is a convex set of actions where the object is involved in the underlying force patterns.

Rosch et al. (1976) point out that some artifacts, such as books, pencils, screwdrivers, and pipe wrenches, have clusters of correlated properties and may therefore be thought of as having some form of essence. This may be accurate, but the set of correlated properties is typically smaller than that for animals. Consequently, the first principal component explains less of the total variance of the concept, which, according to our criterion, indicates a weaker essence. Supporting this view, Gelman (2003, 49) notes that artifacts do not permit inferences to new properties. Additional evidence comes from Nelson (1999), who investigated how toddlers' experiences with object functions affect word learning. She found that toddlers can learn new object names based on function, but only if they had prior direct experience with the objects—in other words, if they had learned something about the actions involved in using the objects.

10.4.3 Essences Change When the Underlying Conceptual Space Changes

A consequence of our characterization of essences as principal components is that judgments of essence can change if the knowledge about a concept changes. This consequence goes against a realist view of essences but is fully compatible with a conceptualist view. Indeed, we think the consequence is desirable. A historical example is the concept WHALE, which used to be categorized as fish. This means that part of its essence was shared with the concept FISH. Nowadays, zoologists tell us that the inner properties of whales entail that they should be categorized as mammals. What has happened is that more physiological domains have been added to the concept WHALE and these additions and their increased salience have drastically changed the principal components of the concept WHALE and thereby its essence.

The pivotal point is that learning concepts does not solely concern grouping features into categories. It also requires understanding the relative significance of these features when categorizing and reasoning with concepts. These kinds of conceptual changes, which appear to be fundamental to children's cognitive development (Carey 2009), have prompted many psychologists to suggest the existence of intuitive theories that organize conceptual knowledge, forming the basis for our judgments of similarity between stimuli (see section 9.1.1). Some psychologists view this as a form of higher-order learning, relating it to the way scientific theories operate (Gelman and Kalish 1993, 110). As mentioned earlier, this suggestion is called the theory-theory.

Scientific theories postulate the existence of unobservable entities and causal structures to account for a range of observable phenomena. If concepts are embedded within theory-like structures, the "superficial" features of their exemplars (such as *shapes*, *colors*, and *sizes*) would be perceived as being causally dependent on unobservable internal features. Furthermore, changes at the level of these causal relationships would generate changes in the category. From our perspective, essentialism is explained quite naturally from our intuitions about the relevance of the internal features of a concept with respect to its external (observable) structure (Gelman and Wellman 1991).

Beyond the objections raised against theory-theory in section 9.1.1, we think that postulating the existence of intuitive theories is less cognitively basic than the extraction of principal components that we suggest as an explanation of how people reason with concepts. Furthermore, the central purpose of theories is to determine the causal relationships underlying domains of phenomena; yet it seems quite hard to explain how children could (implicitly) develop intuitions about causal relations involving features of objects that they cannot perceive. For example, when children think that internal features of animals are essential, they do not rely on biological facts or theories but express their expectations that these features are immutable (Gelman 2003, 240). Our approach avoids these kinds of obscure assumptions as it explains intuitions about essences directly from the detection of stable covariance patterns among observable features. For these reasons, applying Occam's razor, we do not accept theory-theory explanations of essences.

10.5 Essences and Causation

The human mind seeks causes. This is part of what Kant called our *Anschauungsformen (Kant 1997)*. We are causal determinists in the sense that we seek causes also for random events like winning a lottery or becoming sick. Children are also causal determinists, actively expecting events to have causes. Authors such as Rehder (2007, 2017) consider that this bias toward the search for causes plays a central role in explaining essentialism. At first glance, this idea seems to contradict our proposal. However, in what follows, we will show that this is not necessarily the case, and we will discuss how principal components can serve as the causal variables we desire to explain various types of phenomena.

10.5.1 Essences as Inner Causes

We next turn to arguments, in addition to those given in section 9.5, for why principal components can serve as the causal variables we desire to explain various types of phenomena. In brief, essences are seen as inner causes of many of the external features of an object (Gelman 2003, 116–128). Since causes are considered to be more important than effects when objects are categorized (Gelman 2003, 110), it follows that causal dimensions/domains are more salient than the dimensions/domains that represent the effects. This explains why some inner variables primarily determine the categorization of an object, even if the variables are not observed but just postulated.

For example, in Keil's (1989) studies of the transformations of animal kinds, the inner variables dealing with the living and agency aspects of an animal are the variables that determine a concept. Gelman (2003, 128) writes, "Children may possess a skeletal sense of self-generated cause before they have learned about specific internal mechanisms." We suggest that the child's brain somehow identifies the first principal components of a concept, and it is these components that serve as causal variables, even if the child has no way of describing them.

In animal concepts, children and adults may perceive principal components as generative forces behind animal behavior. Gärdenfors (2007, 2014) suggests that actions are categorized by underlying forces,[6] while Gelman (2003) proposes "energy" as a causal placeholder in essentialist thinking. Force and energy are interdefinable, so the two proposals are closely related.

This force-centric perspective corresponds with force-dynamic thinking in cognitive semantics and causal cognition (Talmy 1988; Gärdenfors 2007; Wolff 2007; Croft 2012; Copley and Wolff 2014) and is central to the event model of Gärdenfors and Warglien (2012, see also Gärdenfors 2014, 2024). We hypothesize that the force or energy domain may be among the first principal components to emerge in animal concepts, aligning with our general hypothesis that essences can be represented by principal components. This potentially explains why children view agency, associated with force or energy, as essential for living organisms, thus connecting the concept of essence in animal categories to broader cognitive theories.

Returning to the STRAWBERRY example, one can ask what causes the pleasant taste of strawberries. Both *color* and *sweetness* are positively correlated to *good taste*, but *ripeness* "screens off" the causal effect of the other two variables (Salmon 1984). This means that even though *color* and *sweetness* are elements of *ripeness*, *ripeness* is seen as blocking the causal influence of these variables. For another example, having unprotected sex may cause AIDS, but the intermediate variable, HIV infection, screens off sex as the direct cause of AIDS. And there are other ways of transmitting the HIV virus. This means that we view the virus as the direct cause and unprotected sex as only an indirect variable. We submit that this mechanism works more generally in judgments of essences so that the principal components screen off other variables in causal relations. For example, *living* and *being an agent* screen off the biological variables contributing to these factors. The principal components, therefore, function as hidden causes.

10.5.2 Are Essences Required for Causal Reasoning?

As a criticism of the view that essences are central for causal reasoning, Strevens (2000) argues that the data concerning children's causal inferences does not necessarily imply that they are essentialists. He distinguishes three different forms of essentialism where the version most relevant here is what he calls pure essentialism (see figure 10.1), which consists of three tenets: (1) children represent essences; (2) they define kinds in terms of essences; and (3) they think essences are causally responsible for the observable properties of kinds.

Strevens compares the pure essentialist position to what he calls the "minimal" hypothesis, which denies all three tenets of pure essentialism (see figure 10.2). Instead, the minimal hypothesis says that "children believe there are causal laws connecting natural kinds and their observable properties, but that they are not committed to any particular view about implementing these laws. In particular, children do not represent essences as underlying the laws" (Strevens 2000, 154).

Our main problem with Strevens's hypothesis is that it goes against principles of cognitive economy. A child that represents all the causal relations between natural kinds and their properties (what Strevens calls *K-laws*) would be much more cognitively burdened than a child that generates the causal connections from the essence of the kind. For example, how can a child combine her knowledge of the feathers, wings, weight, flapping, and so on, of a bird with its flying capacity? It is much more cognitively economical to assume that the essence of a bird is what causes it to fly. Again, the essence functions as an intermediate variable between the observable features of a bird and its flying, which screens off the other features from being causes. Since the principal components are the variables that give the most efficient reduction of the variations of the properties of a kind, it will serve as a cognitively economic way of representing causal relations.

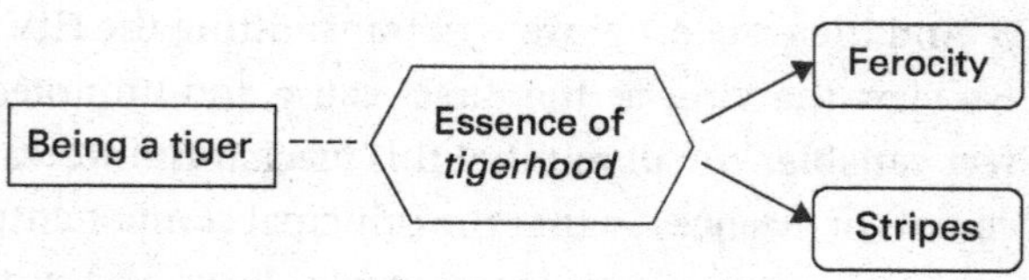

Figure 10.1
Strevens's (2000, 151) depiction of "pure" essentialism.

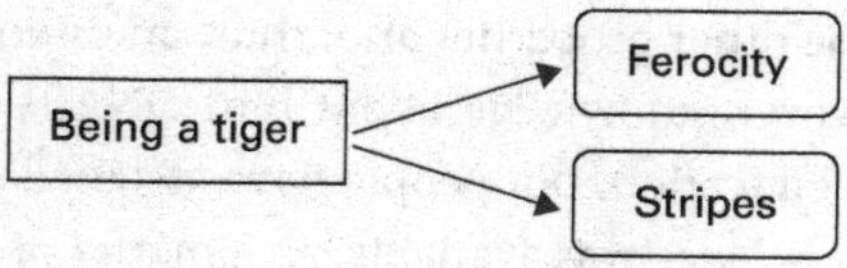

Figure 10.2
Strevens's (2000, 155) depiction of his "minimal" hypothesis.

10.5.3 The Chicken or the Egg?

It might be objected that principal components are not "real" causes but just mathematical constructs. For the die-hard essentialist, essences are "real" and, therefore, cannot be reduced to principal components. Against this view, we have already argued that judgments of essences may vary with the knowledge of the agent, which indicates that essences are, after all, cognitive constructs and nothing inherent in a natural kind. In our opinion, the question of whether principal components are "real" belongs to metaphysics. From our instrumentalist (conceptualist) position, the principal components possess the properties typically associated with causal variables, and they are, therefore, suitable to serve as representations of essences.

Nevertheless, the chicken or the egg problem remains: Which is cognitively prior, essences or causes? In section 9.5, we gave some reasons why covariations are prior to causes when the coherence of concepts is determined. These arguments also apply to essences.

In a line of argumentation that takes causality as the central factor behind essence, Rehder (2007) explored a "generative" theory of concepts that is based on causal networks of properties that belong to a concept (see also Rehder 2017; Rehder and Kim 2006, 2010). For example, bird DNA causes birds to sing, have wings, have light bones, and eat seeds. Having wings and being light causes the bird to be able to fly, and the ability to fly and being light, in turn, is a cause of their building nests in trees (Rehder 2007, figure 12–1). Some of these causal links are probabilistic, and Rehder uses Bayesian calculations to determine the strength of various causal relations. Such generative causal networks are then seen as the basis for the essential properties of a concept. This theory, thus, views causes as cognitively prior to essences.

To construct such causal networks, the relevant causal variable must first be identified. In the example above, "having bird DNA" is seen as the basic cause of all the other properties and, thus, an essential variable of the bird concept. However, knowledge about bird DNA is historically recent and dependent on education, but people have assigned an essence to birds long before this knowledge was available. As a matter of fact, children (and adults) often ascribe essences to natural phenomena without knowing anything explicit about the underlying variables. Therefore, we do not believe that causal models of the kinds studied by Rehder (2007) are helpful in answering the question of whether essences or causes are cognitively more fundamental.

In summary, we believe that causal knowledge is certainly crucial for the perception of the essence of a concept but that this is a phenomenon limited to expert knowledge, whereas, in our everyday knowledge, the perception of essence is based on covariations of properties.

10.6 Essentialism and Language

There are two main connections between essences and language, both of which are central to how children learn concepts. First, when a label is used for a concept, children implicitly assume that the objects that fall under the label share an essence. Given the limited mind-reading of small children (Wellman et al. 2001), it is likely that they believe that labels reflect what objects are called rather than what people call objects. This can be interpreted as a form of essentialist thinking (Gelman 2003). Second, generics are used to inform children about the properties of a concept that help identify the principal components of the concept—that is, its essence.[7]

Having labels for concepts in language has far-reaching consequences. Labels tell children which concepts are kind concepts (see also section 4.7). Naming something ("This is a hedgehog") invites children to form categories. Children implicitly apply the Gricean Maxim of quantity, assuming that novel labels involve rich inferential power. This principle explains why parents normally choose basic-level labels (in the sense of prototype theory) when they introduce a new concept. In support of this, Gelman (2003, 147) writes that children assume "that a property true of one animal is true of other animals in the same basic level category." Linguistically, object concepts are typically referred to by nouns. Carey (1996, 277) says the following about the cognitive work done by nouns: "The child has a default assumption that count nouns are substance sortals . . . the application of every count noun carries with it the idea that the identity of the entity picked out by the noun is unchanged in the face of surface changes." Markman (1989) argues that a noun conveys that a concept (1) supports more inferences, (2) provides more essential information, (3) is central to the identity of an object, (4) is relatively enduring and permanent, (5) is organized into taxonomies, and (6) is nonoverlapping with other concepts. In this way, labeling something with a noun signals that it has an essence.

As we have argued, the principal component carries most of the information about a concept and, therefore, most of the meaning of a noun. Several researchers support this view. Piaget (1929, 61–62) writes that "in learning the names of things, the child at this stage believes it is doing much more. It thinks it is reaching to the essence of the thing and discovering a real explanation." And Carey (1996, 277) submits that essentialism derives from the semantic work done by nouns: "The application of every count noun carries with it the idea that the identity of the entity picked out by the noun is unchanged in the face of surface changes."

In contrast, these features do not hold of adjectives since an adjective only refers to a single domain (Gärdenfors 2014, chap. 7) and can, therefore, neither contain correlations between properties nor be the only components in generics. In agreement with this, Gelman (2003, 187) points out that children "assume that a novel noun refers to a class of like objects, whereas a novel adjective refers to a single property."

10.7 Conclusion

Within philosophy, essentialism has often been associated with a realist view of the world, in particular with so-called modal realism. In this chapter, we have adopted a cognitivist perspective on essences. We believe that such a position can better explain the data that has been collected by psychologists concerning essential thinking in children and adults.

Our contribution is to propose a model for measuring the degree and the content of the essence that people assign to a concept, along with an explanation of how essentialism relates to conceptual coherence. Since the essence of a concept is associated with the covariation of its different dimensions/domains, we have proposed that its essence can be captured by the first principal components. We have demonstrated that this model accounts for several empirical findings on how humans, particularly children, perceive essences. Additionally, we draw connections between essentialist thinking and using labels and generics to convey information about concepts.

While establishing a psychologically valid similarity measure as a basis for distances in conceptual space may be challenging, we believe that our proposal opens up new types of empirical investigations concerning essentialism. In particular, our model predicts that as individuals learn more about a concept, their judgments concerning the essence of the concept will change. This prediction goes against a realist view of essences, and evidence in its favor would support a cognitivist perspective.

11 Conclusion

11.1 Taking Stock

Within analytic philosophy, concept-based reasoning has received little attention, as it conflicts with the traditional assumption that rational inference is grounded on formal-syntactic or probabilistic relations among propositions. Psychology, by contrast, has accumulated an extensive body of theoretical proposals and experimental results on concept-based reasoning. Yet, this research has developed rather fragmentarily, drawing on diverse theoretical frameworks and modeling tools. This lack of a unified perspective has left open fundamental questions about the relationship between reasoning and concepts.

To address this gap, we propose a unifying framework that integrates several areas of research, including category-based induction, nonmonotonic reasoning, generics, analogies, coherence and essentialism. Our proposal is based on the idea that several forms of inference rely on the structure of concepts rather than the syntactic properties of propositions. We have argued that knowledge about such structure is a special type of knowledge—separate from the traditional knowledge-how and knowledge-that—which we call *knowledge-what*. On our account, inductive reasoning aims at creating knowledge-what.

We have proposed conceptual spaces as a general framework for representing knowledge about conceptual structures. The theory of conceptual spaces has already been applied to several other types of cognitive processes, and here we extend it to reasoning.

At the center of our approach are four key measures derived from distances and prototypes in conceptual spaces:

1. *Similarity*, inversely related to distances between points.
2. *Typicality*, measured as distances to prototypes.
3. *Diagnosticity*, defined in terms of typicality and the vertical dimension of the conceptual hierarchy.
4. *Coherence*, determined by covariations of features across domains.

These interrelated measures serve as the foundation for all types of reasoning we have examined. The measure of coherence has also allowed us to present a cognitively grounded account of essentialist thinking with respect to concepts. By incorporating these measures into our models, we have generated novel quantitative predictions that extend beyond the capabilities of previous psychological models (which typically only provide qualitative predictions).

Our framework offers a significant advantage over earlier approaches by providing greater explanatory unification. While other methods focus solely on computational aspects of inference and uncritically assume specific formats for how we mentally representat conceptual information, our framework explains both why concepts have their particular structures and how inferential mechanisms operate over these structures. This comprehensive approach offers deeper insight into the mechanisms underlying human reasoning and categorization.

The models we have presented make it possible to compare the predictions and the results from the different subareas—for example, the relationships between generics and nonmonotonic reasoning based on expectations. The applications of the conceptual space framework to reasoning with concepts extend, however, beyond these topics. The framework can also be applied to related areas such as direct analogies, metaphors, and causal reasoning. Metaphors, in particular, is a central topic within cognitive linguistics, and they have been modeled with the aid of conceptual spaces (Gärdenfors 2000, 2014).

Our approach can be empirically tested through well-established experimental paradigms in psychology. These include induction tasks, which measure subjects' ability to make generalizations based on specific instances of concepts (Heit 1998); feature listing tasks, where participants generate lists of features associated to a concept, with the frequency and order of the features providing insights into the internal structure and organization of the concept (Sloman et al. 1998); and exemplar ratings and typicality judgments, which involve categorizing and rating objects based on their similarity to previous exemplars or to a prototypical instance (Verheyen and Egré 2018). Even computer simulations of categorization and learning can be employed. One such example is Douven and Verheyen (2024), who use simulations of neural networks to show that convexity of conceptual regions facilitates learning more than regions that are merely connected.

Furthermore, spatial arrangement tasks ask participants to physically arrange items based on perceived similarity, offering a spatial representation of conceptual relationships (Richie et al. 2020), and multidimensional scaling uses distance-based statistical techniques on sets of similarity judgments to visually represent the structure of conceptual spaces, allowing for a robust analysis of conceptual similarity and dissimilarity (Rips 1975). Collectively, these diverse paradigms provide a rigorous empirical framework to test the validity and applicability of the models presented in this book.

11.2 The Broader Picture

The theory of conceptual spaces has been proposed as a general framework for knowledge representation in humans (Gärdenfors 2000, 2014). The theory has seen a multitude of applications in several research areas, including philosophy, psychology, computer science, linguistics, and geosciences. In this book we have further extended the applications to the study of reasoning. If fruitfulness is taken as a criterion of success, the theory of conceptual spaces certainly fulfils it.

The plausibility of the theory is strengthened by recent developments in neuroscience, some of which we reported in section 2.8. A central idea is that the hippocampal complex not only represents ordinary space in a coordinate system but also other conceptual features are represented in geometric structures (Bellmund et al. 2018). This area of investigation shows much promise for the future. As we outlined in section 2.9, the theory of conceptual space also opens up for new applications in AI and robotics. Again, this is a research area with great potential.

Despite all these successes, one might ask which alternative theories could compete in representing the structure of human knowledge. In the introductory chapter, we contrasted our proposal with the formalist view of reasoning, particularly with the broader interpretation of this view as a thesis that high-level cognition operates as syntax-driven machinery (Fodor and Pylyshyn 2015). However, if the formalist program is limited to a thesis concerning deductive reasoning—that is, reasoning dependent exclusively on the mastery of logical constants—it may indeed be compatible with our project, since our framework primarily addresses ampliative reasoning. Yet, several other well-developed models of ampliative reasoning exist in psychology and philosophy. Here, we briefly compare the conceptual spaces approach with the two main alternatives: Bayesianism and connectionism (neural networks). Like conceptual spaces, these models also focus on expectation-driven reasoning.

As mentioned in section 1.2, Bayesianism assumes that individuals maintain probability distributions over hypotheses about concepts and update them based on new evidence. This is a very strong ontological assumption, as the underlying hypothesis space in such models is vast, and representing probabilities (and conditional probabilities) over these spaces is highly information demanding. From a psychological perspective, this makes Bayesian models rather implausible, as human memory and attention constraints would severely limit the feasibility of such computations.

Beyond these computational concerns, a deeper line of criticisms targets the explanatory power of Bayesian models in cognitive science. Jones and Love (2011) and Bowers and Davis (2012) argued that Bayesian approaches often rely on post hoc adjustments to priors or likelihoods, which undermines their falsifiability and, ultimately, their predictive power. In line with these points, Endress (2013) showed that Bayesian explanations for rule-learning are difficult to reconcile with standard psychological constraints, as they depend on specialized parameters (e.g., detectors for certain patterns) introduced mainly to align the models with existing data.

A further issue for Bayesian models concerns how they handle the notion of similarity and its role in stimulus generalization. As discussed earlier in this book, conceptual spaces build on Shepard's "universal law of generalization," which assumes the existence of psychological space in which stimuli can be represented as more or less similar to each other based on a distance function. By contrast, Tenenbaum and Griffiths (2001) propose a Bayesian view of generalization based on a "size principle" that holds that rare features should be more relevant for similarity comparisons than common features. Poth (2019) examines whether these two perspectives can be reconciled (see also Poth 2023). Tenenbaum and Griffiths (2001) suggest that "similarity somehow depends on generalization," stating that "the similarity of y to x may involve the probability of generalizing from x to y, or from y to x, or some combination of those two" (2001, 637); but Poth criticizes this way of defining similarity, noting that the size principle itself arguably presupposes a notion of similarity, and that the resulting model risks leading to undergeneralization.

Throughout the book, we have compared our models to existing Bayesian models, and we have claimed that our models have greater explanatory and predictive power. We presume that the debate between researchers advancing probabilistic models and those using geometric models will continue. Since we have proposed several ways of testing the models we have suggested, results from such tests may further inform the debate.

The second main contender for a general account of knowledge representation is connectionism. During the last few decades, the successes of deep learning have been astonishing. It has been shown that these neural networks can learn similarities and can identify prototypes in several types of application areas. Furthermore, they often outperform human classification capabilities. As we have presented in section 2.9, it is possible that the networks generate some kind of (hidden) conceptual spaces that are used in the classifications.

There are, however, some crucial differences between how humans and neural networks handle conceptual information. Some of the spaces are more or less prepared in the human brain—for example, ordinary space, pitch, and color. The mapping from sensory input to the spaces must still be learned, but it is heavily constrained already from the start by the topological organization of the neurons. In contrast, neural networks are typically trained on an initially unstructured network. They must be trained for each application, and they typically do not rely on any underlying topological structure of the "neurons" in the network.[1] Secondly, learning new concepts is, in general, very slow in neural networks and require extensive training, while only a single or a couple of examples may be sufficient for a human to get a good grasp of a novel concept. Unfortunately, is it still to a large extent unknown how this extremely fast learning is accomplished in the brain.

All in all, we believe that the models of reasoning with concepts that we have proposed in this book strengthen the case for conceptual space as the best framework for representing human knowledge structures. We look forward to further testing of the models as part of a general evaluation of the framework.

Notes

Chapter 1

1. Johnson-Laird and Byrne (1992, chap. 3) propose three constraints on rational inference. The first one states that inferences should not "throw away" semantic information. For instance, people do not infer from a premise p, a disjunctive conclusion "p or q" because, even if logically valid, the conclusion is less informative—it is compatible with more possibilities—than the premise. Second, rational inference is parsimonious: We do not conclude, from a set of premises, a conjunction of each of these premises. Again, this is logically valid, yet uninformative. Finally, inferences should add something new to what it is already stated in the premises.

2. For a detailed analysis of this tradition, see MacFarlane (2000).

3. The formalist view of reasoning has played a crucial role in the computer–mind analogy in cognitive science (Boden 1988; Gigerenzer and Goldstein 1996). Roughly put, this influential view holds that cognition is underpinned by formal, mathematical operations that can be modeled algorithmically.

4. In this book, when we refer to logic and logicist models of reasoning, we mean classical logic and its uses in normative and computational models of rational inference (see Tajer 2025). It is important to note, however, that there are many alternative logical systems, several of which were developed specifically to address limitations of classical logic that we highlight throughout this book.

5. Detailed criticisms of the Bayesian program can be found in Bowers and Davis (2012), Knauff and Gazzo Castañeda (2023), and Crupi and Calzavarini (2023).

6. Our approach shares motivations with the theory of mental models (Johnson-Laird 1983). In particular, we agree with the idea that the representational medium for reasoning is likely nonpropositional, or at least not fully so. However, while this theory was proposed as a promising alternative to include the right "machinery to deal with meaning" in reasoning (Byrne 1991), it remains unclear how mental models represent the meaning of lexical concepts that appear in an argument's premises

and conclusion. In other words, the theory lacks a precise account of lexical meaning. Moreover, despite positioning itself as antilogicist, the theory's computational structure and normative framework remain based on logic. The mental models used in reasoning are invariably articulated according to the logical meaning of logical constants, and their conditions of adequacy and validity are also largely logical. The primary distinction between this theory and explicitly logicist theories is its focus on the truth-functional structure of propositions rather than pure syntax. Johnson-Laird's theory is semantic in the same sense that truth tables are (Andrews 1993): It represents possible distributions of truth values for a given proposition based on the distribution of logical constants within it (Lowe 1993; Mercier and Sperber 2017).

Chapter 2

1. There is evidence that large language models represent separable domains as dimensions that are orthogonal to other dimensions (Park et al. 2023).

2. Other notable color spaces include HSV (hue, saturation, value) and CIELab. HSV is similar to NCS but uses "value" instead of "intensity," often visualized as a single cone or cylinder. CIELab, uses L* for lightness and a* and b* for green–red and blue–yellow color components, respectively. CIELab is designed to be perceptually uniform and is often used in color management systems (for a review, see Ibraheem et al. [2012]).

3. An example of how conceptual spaces can accommodate ideas from the enactive approach to concepts can be found in Parthemore (2013). Parthemore extends the original theory by emphasizing that concepts are not static representations but emerge dynamically from agent-environment interactions. It aligns with enactivist perspectives (see Thompson 2010) by arguing that conceptual structures are continuously shaped through engagement with the world, rather than being predefined or solely grounded in neural or symbolic representations.

4. See Kruskal (1964) or Jones and Koehly (1993, 100–101) for a definition of the stress function.

5. Strössner (2022) argues for a weak form of realism.

6. See Schurz (2014, 88–89) for a similar distinction.

7. In chapter 5, we will discuss what is meant by a "natural" concept.

8. Shipley (1993) and Billman and Knutson (1996) showed that covariation between domains in a category has a positive impact in learning. Later in this book, we will show how covariance relates to the coherence of a category.

9. As will be seen in chapter 8, it is important to use covariation rather than correlation.

10. A slightly more complex definition, also involving the *context* where a concept is used, is proposed in chapter 4 of Gärdenfors (2000).

11. It should be noted that this does not necessarily require prototypes being central in the regions they are assigned.

12. Most one-dimensional domains, such as length or temperature, are divided into one lower region (representing "short" or "cold") and one higher (for "tall" or "hot"). None of these regions, however, has a prototype.

13. The assumption that concepts have unique prototypes is an idealization that we use here for the sake of simplicity. A more psychologically realistic model should include the possibility of concepts and properties having multiple prototypes or none (see Douven et al. 2013; Regier et al. 2005).

14. The figure is drawn in two dimensions, but it should be noted that Voronoi tessellations function for any number of dimensions.

15. Also, Johannesson (2000, 2002) develops a "relative prominence model" and performs some experiments that he compares to the results in Nosofsky (1991). Johannesson's model proposes that the experienced directed similarity from x to y is proportional to some symmetric similarity measure between x and y and the quotient between the salience of y and x.

16. There are other possibilities of defining the typicality of properties, such as by using average distances between a prototype and a region. It is a matter of empirical research to determine which method gives the results that best fit with how humans reason.

17. An example of such a model is Lieto and Pozzato (2018).

18. A notion closely related to diagnosticity and widely used in the categorization literature is *cue validity*. While diagnosticity measures how well a property distinguishes one concept from others in its contrast set, cue validity assesses how reliably a trait predicts category membership for a given object. Crucially, cue validity is a probabilistic notion defined as the conditional probability that object o falls in category M given that it possesses feature i: $P(M \mid feature_i)$.

19. Within the field of cognitive and computational linguistics, conceptual spaces have been used to analyze spatial language, the semantics of events and actions (Gärdenfors 2007; Zwarts and Gärdenfors 2016), and other complex linguistic phenomena such as compositionality and conceptual combination (Lewis and Lawry 2016; Bolt et al. 2019). They have also found diverse applications in cognitive modeling. For example, they have been used to model concept creation (McGregor et al. 2015), analogical reasoning (Schockaert and Prade 2013; Douven et al. 2024), musical perception and creativity (Forth et al. 2010; Chella 2015), the representation of pain (Coninx 2022), and human communication (Warglien and Gärdenfors 2013; Wheeler

et al. 2022). In addition, they have found diverse applications in robotics (Chella et al. 2001). Applications in the field of philosophy have also been numerous. For example, the theory has been utilized for the analysis of vagueness (Douven et al. 2013; Mormann 2022), to examine the relationship between perception and belief (Brössel 2017), to represent moral concepts (Verheyen and Peterson 2021), and to address multiple problems in the philosophy of science (Zenker and Gärdenfors 2015; Sznajder 2016; Masterton et al. 2017; Cevolani 2020; De Benedetto 2022; Belastegui 2025).

20. Another application of convexity is Benakkouche et al. (2023), who develops polar convex programming that generates new methods for nonlinear optimization.

21. ChatGPT did not allow the full input of mammal species, so it could not be tested.

Chapter 3

1. Throughout this book, we use the term "uncertainty" to refer specifically to "subjective uncertainty." That is, not uncertainty arising from perceived randomness in a process external to the subject, but rather uncertainty experienced by an agent with respect to his or her own mental processes and representations (Ülkümen et al. 2016). In general, we take uncertainty as an "epistemic feeling"—that is, as an affective experience related to the monitoring of some cognitive process (Proust 2015).

2. Wilfrid Sellars (1953, 1974) was one of the few philosophers who identified this limitation of the formalist tradition and addressed it. According to him, most of our inferential activity is based on the use of "material rules of inference" that operate at the subsentential level by exploiting semantic relations between the nonlogical terms of a natural language.

3. In an *extensional* approach, meaning is specified in terms of extensions or truth-conditions: The set of objects an expression denotes, or the conditions under which a sentence is true. This contrasts with an *intensional* perspective, which characterizes meaning in terms of *senses*, concepts, or possible-worlds, thereby going beyond truth-conditions alone (see Jaquette 2010, chap. 5).

4. Some theories of concepts also assume that attentional mechanisms lie behind the variety of usages that concepts have in everyday cognition. For instance, selective attention seems to play a crucial role in choosing from a concept the relevant information—or features—used in different cognitive procedures, like inference or categorization (Schyns et al. 1998; Barsalou 2003).

5. We use "vector" as "sequence of coordinates." Talking about vectors in a more technical sense could lead to confusion since the dimensions that compose a conceptual space are not necessarily isomorphic to the real numbers; for example, some can be discrete.

6. In naive set theory: For all sets N and M, *if* $x \in N$ *and* $N \subseteq M$, *then* $x \in M$.

7. As Zwarts and Gärdenfors (2016, 118) explain, this is a strong idealization. Landmarks can have different forms, and this can have an impact on the representation of the spatial relations.

Chapter 4

1. Several sections of this chapter build on Gärdenfors and Stephens (2018). We are grateful to Andreas Stephens for the permission to include the material here.

2. In philosophy of science, *induction* typically refers to the process of inferring general laws from the observation of a limited number of particular instances (Schurz 2014, section 2.7). In cognitive psychology, however, the term is often used more broadly, encompassing not only this type of inference but also most cases of reasoning under uncertainty, where an agent relies on background knowledge to make predictions about new objects or situations (Hayes 2007).

3. In the literature, there have been attempts to distinguish "law-like" (nomologic, nomothetic) generalizations from "accidental" generalizations (Goodman 1955; Hempel 1965), and early steps to break the logical emphasis were taken by, for example, Dretske (1977), Tooley (1977) and Armstrong (1978, 1983), who focused on laws as relations of nonlogical *necessitation* between universals (see also Carroll 2020).

4. It should be noted that our use of "knowledge-what" is not intended to cover all everyday uses of the term such as in "I know what time it is."

5. We need not subscribe to full-blown essentialism. It is sufficient that certain properties of spiders are considered cognitively more important than others. See chapter 10 and Gärdenfors (2000, section 4.2.2) for a defense of such a "cognitive" form of essentialism.

6. In different articles, Tulving vacillates in his characterization of memories of facts. Here we link factual knowledge to episodic, rather than to semantic, memory (see also Yee et al. 2014).

7. Kornblith (1993, 105–106) writes, "It is thus safe to say that we have a sensitivity to the features of objects which reside in homeostatic clusters. Indeed, the way in which we detect covariation is precisely tailored to the structure of natural kinds. [. . . We] conceptualize kinds in such a way in order to separate the properties of the members of a kind which are projectable from those which are not. We are aided in this task by our ability to detect clustered covariation."

8. Goodman (1955) spells the term "projectible" in his well-known discussion of induction and entrenched predicates.

Chapter 5

1. Sperber and Wilson (1986) replace this principle by a condition about maximal relevance.

2. Maybe also that it hides its head in the sand (which is not true of ostriches).

3. Formally, if $p_{M,i} \in R_k \subseteq D_i$, then it is expected that $x_i \in R_k \subseteq D_i$.

4. There are other possibilities to define the expectation ordering between properties—for example, by using average distances between a prototype and a region. It is a matter of empirical research to determine which method gives the results that best fit with how humans reason.

5. Since the consequence relation α |~β, denoting that β is a nonmonotonic consequence of α, is equivalent to $\alpha \vdash \beta$ or $\alpha \rightarrow \beta > \alpha \rightarrow \neg\beta$ (see Gärdenfors and Makinson, 1994, 124), the fact that $F(b)$ |~$G(b)$ can be directly proved. This means that an expectation-based inference relation is a nonmonotonic relation with embedded defaults. Thus, we don't need any explicit representation of them.

6. Another example illustrating this point comes from Estes and Glucksberg (1999), who found that people take longer to verify statements like "Peeled apples are round," which combine a typical feature with an atypical modifier. In contrast, they process more quickly statements like "Peeled apples are white," where the property is specifically relevant to the modified subclass. This suggests that we expect modified generics to provide information particularly pertinent to the specified subgroup.

7. Lewis and Lawry (2016) and Lieto and Pozzato (2020) also analyze the conjunction fallacy with the aid of their respective models of concept combination, so it is interesting to compare with their solutions. The goal of the model presented by Lewis and Lawry (2016) is to represent hierarchies of concepts. Their model is similar to ours in that they also use a geometrical approach based on conceptual spaces. Instead of using Voronoi tessellations to determine category membership, they use random set theory. Nevertheless, since their explanation of the conjunction fallacy is also based on distances to prototypes, it is similar in spirit to ours. The description logic presented by Lieto and Pozzato (2020) is less similar since their representations are probabilistic rather than geometric. Therefore, it is difficult to compare their account of the conjunction fallacy to the solution we present. There are, nevertheless, interesting similarities between the models: The logic for their typicality operator can be shown to be equivalent to the nonmonotonic logic presented by Gärdenfors and Makinson (1994).

8. A related "fallacy" is the disjunction fallacy, which refers to the tendency for people to judge a specific concept as more probable than a broader, superordinate concept that includes it, violating the rule that the probability of a disjunction (A

or B) must be greater than or equal to the probability of either of its constituents (A, B). For instance, in a study by Bar Hillel and Neter (1993), participants often rated a basic-level concept like "tennis" for an athletic, posh, and vain person as more likely than its superordinate concept, "ball game," even though the superordinate concept must be at least as probable ("ball game" can be seen as a disjunction of every ball game, including tennis).

9. For instance, one could argue that the inferences "A entails A" and "If it rains, therefore A entails A" are less informative than "If it rains, the streets are wet. The streets are not wet; therefore, it does not rain," since the first is trivial and the second involves entirely unrelated topics while the third articulates coherently meaningful information.

10. For example, Lakoff (1971: 133) distinguishes between two main meanings of "but": (1) semantic opposition as in "John is tall but Bill is short"; and (2) denial of expectation as in "John is tall but he's no good at basketball." However, as we will argue later, the semantic opposition meaning is a special case of violated expectation.

Chapter 6

1. Prasada et al. (2013) distinguish between majority characteristic, minority characteristic, majority statistical, striking, and quasi-definitional generics.

2. In section 5.4, we described three different ways in which our expectations about object-properties are updated in the light of new information.

3. The previous explanations cohere with Cimpian and Markman's (2009) hypothesis that generics about biological kinds are interpreted in the light of naïve biological theories implicitly held by cognitive agents.

4. Cohen (2001, 60–64) makes a distinction between "absolute" and "relative" generics that is seemingly similar to ours. A main difference, however, is that he analyzes generics in terms of probabilities rather than our use of typicality and diagnosticity. Leslie and Learner (2016, section 4) argue that there are cases that Cohen's probabilistic approach cannot handle.

5. Children's picture books of animals and other object concepts highlight the diagnostic properties of the concepts.

6. Van Rooij and Schulz (2020) emphasize the role of learning generics.

Chapter 7

1. Following the terminology presented in section 2.4, we will use the term "object concept" (or just "concept") instead of "category."

2. Typicality is strongly related to similarity (see Rips 1989) as is witnessed by our typicality measure.

3. For a discussion of the role of prototypicality in reasoning in general, see Cherniak (1984).

4. Note that here we refer to the typicality of a concept with respect to other concepts in its contrast class and not to the typicality of a property, as in the case of our typicality criterion.

5. An alternative idea is to explicitly introduce distances between regions as a function of distances between their points. It would be a matter of empirical testing to determine which model would give the best results.

6. It is possible that a concept with greater volume is less typical than a concept with a smaller volume. For example, FISH may have a greater volume than CAT but be less typical as a PET. However, in equation 7.3, the expectation value is not only determined by the volume of a concept but also its typicality. So even though FISH may have a larger volume than CAT, the greater typicality of CAT will counterweigh this.

7. A reason for this assignment is that if Y is a region that partially overlaps X and then shrinks to become a subset of X, then $V(Y-X)$ will be smaller and smaller, which means that $ExpS(X \rightarrow Y)_Z$ will approach infinity. From a psychological perspective, we hypothesize that these cases require subjects to use an inferential mechanism based on class inclusion, like property inheritance. If $Y \subset X$, members of Y inherit all properties of X, and thus $ExpS(X \rightarrow Y)_Z$ is maximal.

8. For cases in which this condition does not hold, it is possible that the volume of the hull of two large contiguous regions is larger than the hull of two distant small regions. An empirical study of this fact could be a way of testing the fruitfulness of the notion of volume of a category for the analysis of CBI.

9. This assumption is not meant to be psychologically realistic. Prototypes are hardly centroids of the convex regions that represent them, even for natural concepts (Douven 2019). However, according to the empirical literature, the typicality effect holds for multiple-premise arguments as a compound measure of the degree of typicality of some of the concepts in the premises. Considering the lack of robust evidence about how these degrees of typicality interact, we introduced the centrality of the artificial prototype as a formal idealization that seems to respond well to the classical examples.

10. One area that we will not consider in this chapter is the influence of causal relations between the concepts involved. Various experimental studies have shown that causal knowledge is important in CBI, sometimes overriding standard similarity and typicality relations (Medin et al. 2003; Rehder and Hastie 2001; Shafto et al. 2007; Bright and Feeney 2014). For example, "Grass has enzyme E; thus, cows have

enzyme E" is judged to be stronger than "Cows have enzyme E; thus, grass has enzyme E" since there may exist a causal link from the enzyme of the grass to the enzyme of the cows. One possible way to use our model also for these phenomena is that causal connections may introduce a different kind of "typicality" relations between the concepts so that the presence of the enzyme is more typical for grass than for cows.

11. $F(X)\ F(Y)$ is the inner product of the two vectors, defined as $\sum_i F(X)_i \cdot F(Y)_i$. $|F(Y)|^2$ is the inner product of $F(Y)$ with itself, defined as $\sum_i F(Y)_i^2$.

12. See Rogers and McClelland (2004) for a connectionist approach to semantic cognition.

13. Also Rips (1975) uses data from MDS for analyzing CBI arguments.

Chapter 8

1. For notational simplicity, we will use capital letters for all the terms in an analogy, regardless of whether they are concepts, properties, or verbs.

2. In computer science and AI, a "search problem" is a type of task that requires search algorithms to explore a problem space and find optimal solutions or paths. These algorithms are used for optimization, decision-making, and problem-solving in various domains.

3. Gentner (1983) defines relational predicates as n-ary predicates with $n > 1$.

4. Notice that M and M^* are the same search space since they include the same set of subcategories.

5. Alternatively, the skeleton can be defined as the sets of centers of discs that "touch" the shape's boundary in at least two points.

6. Note that if the order of terms in the analogy is *A:B::C:X* such that *A* and *C* are parts of *B* and *X*, respectively, the search procedure will be different. In this case, it will be necessary to specify a search space *S* of lexical categories such that $S = \{Y: Y$ satisfies the condition *has_part_C*$\}$. Then, Zhu and Yuille's algorithm can be applied to find the most similar skeletal shape to *B* in *S* such that part *A* in *B* is mapped onto part *C* in *X*. Since it is very likely that there are various concepts in S that are good candidates for completing *A:B::C:X*, the above algorithm can be used to find a set of skeletal shapes in *S* that meet or exceed a similarity threshold with regard to *B*.

7. A fifth type, which we do not study in this chapter, are functional (causal) relations—for example, components such as DOG:LEASH::STUDENT:NOTEBOOK or composed analogies such as HUMAN:BRAIN::COMPUTER:CPU. For some ideas about how functions can be analyzed with the aid of conceptual spaces, see Gärdenfors (2007).

Chapter 9

1. This mirrors what researchers in machine learning call the "curse of dimensionality" (Bellman 1961): As the number of relevant dimensions increases, the amount of information required to represent and generalize grows exponentially.

2. Numerous probabilistic models of coherence have been developed within the realm of formal epistemology (see, e.g., Douven and Meijs [2007] and Hartmann and Trpin [2023]). However, these models address a phenomenon distinct from our primary interest here. Specifically, they aim to measure the coherence between sets of propositions, rather than the coherence of the internal structure of concepts.

3. It is important to note that the idea that our cognitive system operates by compressing and simplifying information is not new. Edelman and Intrator (1997) have argued that one of the central functions of learning is to extract low-dimensional representations from high-dimensional input spaces. According to them, perceptual and conceptual learning alike can be understood as processes of dimensionality reduction, in which the system identifies a small set of latent variables that explain most of the variation in the data, thereby enabling both efficient storage and generalization. Similarly, recent research on episodic memory suggests that dimensionality reduction may play a role in memory retrieval. Studies by Kerrén and colleagues (2025) suggest that the ability to encode and retrieve episodes depends on compressing high-dimensional experiential information into a more compact representational format. This compression is thought to reduce interference among stored memories while preserving the relational structure needed for flexible retrieval.

4. Rogers and McClelland (2004, chap. 9) propose a similar argument, supported by simulations in their Parallel Distributed Processing model. They show that the model learns to weight properties by estimating not only the covariances observed within a single concept, but also the covariances that hold between concepts across an entire domain. For example, the network discovers that *lives near water* and *eats fish* co-occur not just within a specific group of birds, but across multiple animal domains (fish, amphibians, shore mammals). Because this *habitat–diet* link is consistent across different domains, the correlation at the individual concept level appears stronger. Conversely, a pairing like *is white* and *eats fish*, which lacks this kind of cross-domain regularity, receives a lower weight and consequently appears weaker.

5. Inference learning is often contrasted with "classification learning," which consists of predicting a concept label by identifying diagnostic features of exemplars.

6. Siebel follows Bonjour in this regard: The coherence of a system of beliefs is diminished to the extent to which it is divided into subsystems of beliefs that are relatively unconnected to each other by inferential connections (Bonjour 1985, 98).

Chapter 10

1. This expression was introduced by Medin and Ortony (1989).

2. A related problem is how to distinguish between "kinds" without essences and those with (Noyes and Keil 2019; Vasilyeva and Lombrozo 2020). The notion of "kind" is not central, so we refrain from describing the results.

3. An exception is Rehder (2007). We return to his model in section 10.5.

4. This does not mean that only concepts of which we have deep knowledge are "essentialized." Ahn et al. (2013) demonstrated that merely perceiving a concept as a kind is sufficient for people to assume it has an essence. We think that, in these cases, people make a meta-inference based on prior knowledge, assuming that the essentialized concept possesses a covariate feature structure even when they lack specific knowledge of it.

5. Insect-eating plants are often considered strange since they seem to exhibit some form of agency.

6. Forces are not just physical but also include, for example, mental, chemical, and medical factors.

7. That these two linguistic categories are central is supported by Gelman (2003, 324), who wrote that "language is implicated in two sorts of ways. First, labels and generics tell children when (which kinds) to essentialize. . . . Second, some of the capacities required for language learning appear also to serve essentializing functions: namely the logic of sortals . . . and the logic of proper names."

Chapter 11

1. Exceptions are Kohonen's self-organizing maps (Kohonen 1990) and growing grids, which come close to the methodology of conceptual spaces (Gärdenfors 2000).

References

Abdi, Hervé, and Lynne J. Williams. 2010. Principal component analysis. *Computational Statistics* 2 (4): 433–459.

Adams, Benjamin, and Martin Raubal. 2009. Conceptual space markup language (CSML): Towards the cognitive semantic web. In *IEEE International Conference on Semantic Computing*, 253–260. IEEE.

Ahlqvist, Ola. 2005. Using uncertain conceptual spaces to translate between land cover categories. *International Journal of Geographical Information Science* 19 (7): 831–857.

Ahn, Woo-kyoung, Eric G. Taylor, Daniel Kato, Jessecae K. Marsh, and Paul Bloom. 2013. Causal essentialism in kinds. *Quarterly Journal of Experimental Psychology* 66 (6): 1113–1130.

Ahn, Woo-kyoung, Jessecae K. Marsh, Christian C. Luhmann, and Kevin Lee. 2002. Effect of theory-based feature correlations on typicality judgments. *Memory & Cognition* 30: 107–118.

Aldridge, Susan, John Parascandola, and Jeffrey L. Sturchio. 1999. *Discovery and Development of Penicillin*. American Chemical Society.

Andrews, Avery D. 1993. Mental models and tableau logic. *Behavioral and Brain Sciences* 16 (2): 334–334.

Aristotle. 1965. *Prior and Posterior Analytics*. Edited by W. D. Ross. Oxford University Press.

Armstrong, David M. 1978. *A Theory of Universals*. Cambridge University Press.

Armstrong, David M. 1983. *What Is a Law of Nature?* Cambridge University Press.

Aronov, Dmitriy, Rhino Nevers, and David W. Tank. 2017. Mapping of a non-spatial dimension by the hippocampal–entorhinal circuit. *Nature* 543 (7647): 719–722.

Barbot, Nelly, Laurent Miclet, and Henri Prade. 2019. Analogy between concepts. *Artificial Intelligence* 275:487–539.

Barsalou, Lawrence W. 1983. Ad hoc categories. *Memory and Cognition* 11:211–227.

Barsalou, Lawrence. 1987. The instability of graded structure: Implications for the nature of concepts. In *Concepts and Conceptual Development: Ecological and Intellectual Factors in Categorization*, edited by Ulric Neisser, 101–140. Cambridge University Press.

Barsalou, Lawrence. 2003. Situated simulation in the human conceptual system. *Language and Cognitive Processes* 18 (5–6): 513–562.

Behrens, Heike. 2017. The role of analogy in language processing and acquisition. In *The Changing English Language: Psycholinguistic Perspectives*, edited by Marianne Hundt, Sandra Mollin, and Simone E. Pfenninger, 215–239. Cambridge University Press.

Bellmund, Jacob L. S., Peter Gärdenfors, Edvard I. Moser, and Christian F. Doeller. 2018. Navigating cognition: Spatial codes for human thinking. *Science* 362 (6415): eaat6766.

Benakkouche, Lilia, Blake Whitman, and Baha Alzalg. 2023. Polar convex programming: A new paradigm for nonlinear optimization. *Applied Mathematics* 17 (3): 469–481.

Belastegui, Javier. 2025. A conceptual space for classical concepts. *Synthese* 206 (2): 1–47.

Bellman, Richard. 1961. *Adaptive Control Processes: A Guided Tour*. Princeton University Press.

Berlincourt, Ted Gibbs, and Richard R. Hake. 1963. Superconductivity at high magnetic fields. *Physical Review* 131 (1): 140–157.

Billman, Dorrit. 1983. *Procedures for Learning Syntactic Structure: A Model and Test with Artificial Grammars*. Doctoral dissertation, University of Michigan.

Billman, Dorrit, and David Dávila. 2001. Consistent contrast aids concept learning. *Memory and Cognition* 29 (7): 1022–1035.

Billman, Dorrit, and James Knutson. 1996. Unsupervised concept learning and value systematicity: A complex whole aids learning the parts. *Journal of Experimental Psychology: Learning, Memory, and Cognition* 22 (2): 458–475.

Binder, Jeffrey R., and Rutvik H. Desai. 2011. The neurobiology of semantic memory. *Trends in Cognitive Sciences* 15 (11): 527–536.

Birnbaum, Michael H., Carolyn J. Anderson, and Unda G. Hynan. 1990. Theories of bias in probability judgment. *Advances in Psychology* 68: 477–498.

Boden, Margaret A. 1988. *Computer Models of Mind: Computational Approaches in Theoretical Psychology*. Cambridge University Press.

Bolt, Joe, Bob Coecke, Fabrizio Genovese, Martha Lewis, Dan Marsden, and Robin Piedeleu. 2019. Interacting conceptual spaces I: Grammatical composition of concepts. In *Conceptual Spaces: Elaborations and Applications*, edited by Mauri Kaipainen, Frank Zenker, Antti Hautamäki, and Peter Gärdenfors, 151–181. Springer.

BonJour, Laurence. 1985. *The Structure of Empirical Knowledge*. Harvard University Press.

Borghi, Anna M., Ferdinand Binkofski, Cristiano Castelfranchi, Felice Cimatti, Claudia Scorolli, and Luca Tummolini. 2017. The challenge of abstract concepts. *Psychological Bulletin* 143 (3): 263–292.

Bouraoui, Zied, Víctor Gutiérrez-Basulto, and Steven Schockaert. 2022. Integrating ontologies and vector space embeddings using conceptual spaces (Invited Paper). Presented at the International Research School in Artificial Intelligence, Bergen, Norway.

Bovens, Luc, and Stephan Hartmann. 2003. Solving the riddle of coherence. *Mind* 112 (448): 601–633.

Bowers, Jeffrey S., and Colin J. Davis. 2012. Bayesian just-so stories in psychology and neuroscience. *Psychological Bulletin* 138 (3): 389–414.

Brachman, Ronald J. 1979. On the epistemological status of semantic networks. In *Associative Networks*, edited by Nicholas Findler, 3–50. Academic Press.

Braine, Martin D. S. 1994. Mental logic and how to discover it. In *The Logical Foundations of Cognition*, edited by John Macnamara and Gonzalo E. Reyes, 241–263. Oxford University Press.

Brandom, Robert. 2001. *Articulating Reasons: An Introduction to Inferentialism*. Harvard University Press.

Brandone, Amanda C. 2017. Changes in beliefs about category homogeneity and variability across childhood. *Child Development* 88 (3): 846–866.

Brewka, Gerhard, Floris Roelofsen, and Luciano Serafini. 2007. Contextual default reasoning. In *Proceedings of the 20th International Joint Conference on Artificial Intelligence (IJCAI'07)*, edited by Rajeev Sangal, Harish Mehta, and R. K. Bagga, 268–273. Morgan Kaufmann.

Briggs, Stephen R., and Jonathan M. Cheek. 1986. The role of factor analysis in the development and evaluation of personality scales. *Journal of Personality* 54 (1): 106–148.

Bright, Aimée K., and Aidan Feeney. 2014. Causal knowledge and the development of inductive reasoning. *Journal of Experimental Child Psychology* 122: 48–61.

Brössel, Peter. 2017. Rational relations between perception and belief: The case of color. *Review of Philosophy and Psychology* 8: 721–741.

Buckner, Cameron. 2019. Rational inference: The lowest bounds. *Philosophy and Phenomenological Research* 98 (3): 697–724.

Byrne, Ruth M. J. 1991. Can valid inferences be suppressed? *Cognition* 39 (1): 71–78.

Caglar, Leyla R. 2021. *Conceptual Spaces in the Brain: An Exploration of Structure Shape and Categorization*. Doctoral dissertation, Rutgers University.

Cann, Ronnie. 1993. *Formal Semantics: An Introduction*. Cambridge University Press.

Carey, Susan. 1985. *Conceptual Change in Childhood*. MIT Press.

Carey, Susan. 1996. On the origin of causal understanding. In *Causal Cognition: A Multidisciplinary Debate*, edited by Dan Sperber, David Premack, and Ann James Premack, 268–308. Oxford University Press.

Carey, Susan. 2009. *The Origin of Concepts*. Oxford University Press.

Carlson, Gregory N. 2009. Generics and concepts. In *Kinds, Things and Stuff: Mass Terms and Generics*, edited by Francis Jeffry Pelletier, 16–36. Oxford University Press.

Carnap, Rudolf. 1950. *Logical Foundations of Probability*. Chicago University Press.

Carnap, Rudolf. 1952. Meaning postulates. *Philosophical Studies* 3: 65–73.

Carnap, Rudolf. 1971. A basic system of inductive logic, part 1. In *Studies in Inductive Logics and Probability*, edited by Rudolf Carnap and Richard C. Jeffrey, vol. 1, 35–165. University of California Press.

Carroll, John W. 2020. Laws of nature. Stanford Encyclopedia of Philosophy Archive (Winter 2020 Edition), edited by Edward N. Zalta. https://plato.stanford.edu/archives/win2020/entries/laws-of-nature/.

Cevolani, Gustavo. 2020. Approaching truth in conceptual spaces. *Erkenntnis* 85 (6): 1485–1500.

Chater, Nick, and Paul Vitányi. 2003. Simplicity: A unifying principle in cognitive science? *Trends in Cognitive Sciences* 7 (1): 19–22.

Chatterjee, Usashi, Amit Gajbhiye, and Steven Schockaert. 2023. Cabbage sweeter than cake? Analysing the potential of large language models for learning conceptual spaces. arXiv preprint arXiv:2310.05481.

Chatterjee, Usashi, and Steven Schockaert. 2023. Probing the conceptual space of ChatGPT and GPT-4. Presented at the 9th Workshop on Artificial Intelligence and Cognition, Bremen, 14–15.

Chella, Antonio. 2015. A cognitive architecture for music perception exploiting conceptual spaces. In *Applications of Conceptual Spaces: The Case for Geometric Knowledge Representation*, edited by Frank Zenker and Peter Gärdenfors, 187–203. Springer International.

Chella, Antonio, Massimo Frixione, and Salvatore Gaglio. 2001. Conceptual spaces for computer vision representations. *Artificial Intelligence Review* 16: 137–152.

Cherniak, Christopher. 1984. Prototypicality and deductive reasoning. *Journal of Verbal Learning and Verbal Behavior* 23 (5): 625–642.

Chierchia, Gennaro. 1995. Individual level predicates as inherent generics. In *The Generic Book*, edited by Francis Jeffry Pelletier and Gregory Carlson, 176–223. Chicago University Press.

Chierchia, Gennaro, and Sally McConnell-Ginet. 1990. *Meaning and Grammar: An Introduction to Semantics*. MIT Press.

Chin-Parker, Seth, and Brian H. Ross. 2004. Diagnosticity and prototypicality in category learning: A comparison of inference learning and classification learning. *Journal of Experimental Psychology: Learning, Memory, and Cognition* 30 (1): 216–226.

Cimpian, Andrei, and Ellen M. Markman. 2009. Information learned from generic language becomes central to children's biological concepts: Evidence from their open-ended explanations. *Cognition* 113 (1): 14–25.

Cimpian, Andrei, and Rose M. Scott. 2012. Children expect generic knowledge to be widely shared. *Cognition* 123 (3): 419–433.

Cimpian, Andrei, Susan A. Gelman, and Amanda C. Brandone. 2010. Theory-based considerations influence the interpretation of generic sentences. *Language and Cognitive Processes* 25 (2): 261–276.

Clark, Eve V. 1993. *The Lexicon in Acquisition*. Cambridge University Press.

Cohen, Ariel. 2001. Relative readings of many, often, and generics. *Natural Language Semantics* 9:41–67.

Cohen, Ariel. 2004. Generics and mental representations. *Linguistics and Philosophy* 27: 529–556.

Coley, John D., and Nadya Vasilyeva. 2010. Generating inductive inferences: Premise relations and property effects. *Psychology of Learning and Motivation* 53: 183–226.

Coley, John D., Patrick Shafto, Olga Stepanova, and Elizabeth Baraff. 2005. Knowledge and category-based induction. In *Categorization Inside and Outside the Laboratory: Essays in Honor of Douglas L. Medin*, edited by Woo-kyoung Ahn, Robert L. Goldstone, Bradley C. Love, Arthur B. Markman, and Phillip Wolff, 69–85. American Psychological Association.

Coninx, Sabrina. 2022. A multidimensional phenomenal space for pain: Structure, primitiveness, and utility. *Phenomenology and the Cognitive Sciences* 21 (1): 223–243.

Connolly, Andrew C., Jerry A. Fodor, Lila R. Gleitman, and Henry Gleitman. 2007. Why stereotypes don't even make good defaults. *Cognition* 103 (1): 1–22.

Constantinescu, Alexandra O., Jill X. O'Reilly, and Timothy E. J. Behrens. 2016. Organizing conceptual knowledge in humans with a gridlike code. *Science* 352 (6292): 1464–1468.

Cooper, Gregory F. 1990. The computational complexity of probabilistic inference using Bayesian belief networks. *Artificial Intelligence* 42 (2–3): 393–405.

Copley, Bridget, and Phillip Wolff. 2014. Theories of causation should inform linguistic theory and vice versa. In *Causation in Grammatical Structures*, edited by Bridget Copley and Fabienne Martin, 10–57. Oxford University Press.

Corcoran, John. 2006. Schemata: The concept of schema in the history of logic. *Bulletin of Symbolic Logic* 12 (2): 219–240.

Corter, James E., and Mark A. Gluck. 1992. Explaining basic categories: Feature predictability and information. *Psychological Bulletin* 111 (2): 291-303.

Costello, Fintan J. 2009. How probability theory explains the conjunction fallacy. *Journal of Behavioral Decision Making* 22 (3): 213–234.

Creath, Richard. 2014. Logical empiricism. In *The Stanford Encyclopedia of Philosophy* (Spring 2014 Edition), edited by Edward N. Zalta.

Croft, William. 2012. *Verbs: Aspect and Causal Structure*. Oxford University Press.

Crupi, Vincenzo, and Fabrizio Calzavarini. 2023. Critique of pure Bayesian cognitive science: A view from the philosophy of science. *European Journal for Philosophy of Science* 13 (28): 1–17. https://doi.org/10.1007/s13194-023-00533-w.

Cruse, D. Alan. 1986. *Lexical Semantics*. Cambridge University Press.

Cruse, D. Alan. 2000. *Meaning in Language: An Introduction to Semantics and* Pragmatics. Oxford University Press.

Csibra, Gergely, and György Gergely. 2009. Natural pedagogy. *Trends in Cognitive Sciences* 13 (4): 148–153.

Cubek, Richard, Wolfgang Ertel, and Günther Palm. 2015. High-level learning from demonstration with conceptual spaces and subspace clustering. In *2015 IEEE International Conference on Robotics and Automation (ICRA)*, 2592–2597. IEEE.

Cummings, Louise. 2005. *Pragmatics: A Multidisciplinary Perspective*. Routledge.

Cummings, Louise. 2014. *Reasoning and Public Health: New Ways of Coping with Uncertainty*. Springer.

De Benedetto, Matteo. 2022. Explicating "explication" via conceptual spaces. *Erkenntnis* 87 (2): 853–889.

Declerck, Renaat. 1991. The origins of genericity. *Linguistics* 29 (1): 79–102.

Decock, Lieven, and Igor Douven. 2011. Similarity after Goodman. *Review of Philosophy and Psychology* 2 (1): 61–75.

Decock, Lieven, and Igor Douven. 2014. What is graded membership? *Noûs* 48 (4): 653–682.

Delgrande, James P. 2011. What's in a default? In *Nonmonotonic Reasoning: Essays Celebrating its 30th Anniversary*, edited by Gerhard Brewka, Victor W. Marek, and Mirosław Truszczyński, 89–109. College Publications.

Delgrande, James P., and Torsten Schaub. 2000. Expressing preferences in default logic. *Artificial Intelligence* 123 (1–2): 41–87.

Dennett, Daniel C. 1991. Real patterns. *Journal of Philosophy* 88 (1): 27–51.

Devadoss, Satyan, and Joseph O'Rourke. 2011. *Discrete and Computational Geometry*. Princeton University Press.

Dinsmore, John. 2014. Thunder in the gap. In *The Symbolic and Connectionist Paradigms*, edited by John Dinsmore, 1–24. Psychology Press.

Dorszewski, Teresa, Lenka Tětková, Lorenz Linhardt, and Lars Kai Hansen. 2024. Connecting concept convexity and human-machine alignment in deep neural networks. arXiv preprint arXiv:2409.06362.

Douven, Igor. 2016. Vagueness, graded membership, and conceptual spaces. *Cognition* 151: 80–95.

Douven, Igor. 2019. Putting prototypes in place. *Cognition* 193: 104007.

Douven, Igor, Lieven Decock, Richard Dietz, and Paul Égré. 2013. Vagueness: A conceptual spaces approach. *Journal of Philosophical Logic* 42 (1): 137–160.

Douven, Igor, Shira Elqayam, Peter Gärdenfors, and Patricia Mirabile (2022). Conceptual spaces and the strength of similarity-based arguments. *Cognition*, 218, 104951.

Douven, Igor, and Peter Gärdenfors. 2019. What are natural concepts? A design perspective. *Mind and Language* 35 (3): 313–334.

Douven, Igor, and Steven Verheyen. 2024. Concept learning: Convexity versus connectedness. *Erkenntnis*, 1–18. https://doi.org/10.1007/s10670-024-00909-1.

Douven, Igor, Steven Verheyen, Shira Elqayam, Peter Gärdenfors, and Matías Osta-Vélez. 2023. Similarity-based reasoning in conceptual spaces. *Frontiers in Psychology* 14: 1234483.

Douven, Igor, Steven Verheyen, Shira Elqayam, Peter Gärdenfors, and Matías Osta-Vélez. 2025. Analogical reasoning: A Carnapian approach. *Synthese* 205 (185): 1–34. https://doi.org/10.1007/s11229-025-04991-y.

Douven, Igor, and Wouter Meijs. 2007. Measuring coherence. *Synthese* 156: 405–425.

Dretske, Fred. 1977. Laws of nature. *Philosophy of Science* 44: 248–268.

Edelman, Shimon, and Nathan Intrator. 1997. Learning as extraction of low-dimensional representations. *Psychology of Learning and Motivation* 36: 353–380.

Endress, Ansgar D. 2013. Bayesian learning and the psychology of rule induction. *Cognition* 127 (2): 159–176.

Evans, Jonathan St. B. T. 1989. Concepts and inference. *Mind and Language* 4 (1–2): 29–34.

Evans, Jonathan St. B. T. 1991. Theories of human reasoning: The fragmented state of the art. *Theory and Psychology* 1 (1): 83–105.

Evans, Jonathan St. B. T., and Aidan Feeney. 2004. The role of prior belief in reasoning. In *The Nature of Reasoning*, edited by Jacqueline P. Leighton and Robert J. Sternberg, 78–102. Cambridge University Press.

Evans, Jonathan St. B. T., and David E. Over. 2004. *If: Supposition, Pragmatics, and Dual Processes*. Oxford University Press.

Evans, Jonathan St. B. T., Julie L. Barston, and Paul Pollard. 1983. On the conflict between logic and belief in syllogistic reasoning. *Memory and Cognition* 11 (3): 295–306.

Fairbanks, Grant, and Patti Grubb. 1961. A psychophysical investigation of vowel formants. *Journal of Speech and Hearing Research* 4 (3): 203–219.

Fantino, Edmund, James Kulik, Stephanie Stolarz-Fantino, and William Wright. 1997. The conjunction fallacy: A test of averaging hypotheses. *Psychonomic Bulletin and Review* 4: 96–101.

Fantl, Jeremy. 2016. Knowledge how. In *Stanford Encyclopedia of Philosophy*, edited by Edward N. Zalta. https://plato.stanford.edu/archives/spr2021/entries/knowledge-how/.

Feeney, Aidan. 2018. Forty years of progress on category-based inductive reasoning. In *The Routledge International Handbook of Thinking and Reasoning*, edited by Linden J. Ball and Valerie A. Thompson, 167–185. Routledge/Taylor & Francis Group.

Feeney, Aidan, and Evan Heit. 2011. Properties of the diversity effect in category-based inductive reasoning. *Thinking and Reasoning* 17 (2): 156–181.

Fillmore, Charles J. 1982. Frame semantics. In *Linguistics in the Morning Calm*, edited by The Linguistic Society of Korea, 111–137. Hanshin Publishing.

Fillmore, Charles J., and Beryl T. Atkins. 2000. Describing polysemy: The case of "crawl." In *Polysemy: Theoretical and Computational Approaches*, edited by Yael Ravin and Claudia Leacock, 91–110. Oxford University Press.

Fiorini, Sandro R., Peter Gärdenfors, and Mara Abel. 2014. Representing part–whole relations in conceptual spaces. *Cognitive Processing* 15 (2): 127–142.

Firestone, Chaz, and Brian J. Scholl. 2014. "Please tap the shape, anywhere you like": Shape skeletons in human vision revealed by an exceedingly simple measure. *Psychological Science* 25: 377–386.

Fodor, Jerry A. 1975. *The Language of Thought*. Harvard University Press.

Fodor, Jerry A. 1980. Methodological solipsism considered as a research strategy in cognitive psychology. *Behavioral and Brain Sciences* 3 (1): 63–73.

Fodor, Jerry A. 1983. *The Modularity of Mind*. MIT Press.

Fodor, Jerry A. 1985. Fodor's guide to mental representation: The intelligent auntie's vade-mecum. *Mind* 94 (373): 76–100.

Fodor, Jerry A. 1987. *Psychosemantics: The Problem of Meaning in the Philosophy of Mind*. MIT Press.

Fodor, Jerry A., and Zenon W. Pylyshyn. 1988. Connectionism and cognitive architecture: A critical analysis. *Cognition* 28 (1–2): 3–71.

Fodor, Jerry A., and Zenon W. Pylyshyn. 2015. *Minds without Meanings: An Essay on the Content of Concepts*. MIT Press.

Forth, Jamie, Geraint A. Wiggins, and Alex McLean. 2010. Unifying conceptual spaces: Concept formation in musical creative systems. *Minds and Machines* 20: 503–532.

Gärdenfors, Peter. 1988. *Knowledge in Flux: Modeling the Dynamics of Epistemic Systems*. MIT Press.

Gärdenfors, Peter. 1990. Induction, conceptual spaces and AI. *Philosophy of Science* 57 (1): 78–95.

Gärdenfors, Peter. 1994. The role of expectations in reasoning. In *Knowledge Representation and Reasoning Under Uncertainty*, edited by Michael Masuch and László Pólos, 1–16. Springer.

Gärdenfors, Peter. 1997. Symbolic, conceptual and subconceptual representations. In *Human and Machine Perception*, edited by Virginio Cantoni, Vito Di Gesu, Alessandra Setti, and Domenico Tegolo, 255–270. Springer US.

Gärdenfors, Peter. 2000. *Conceptual Spaces: The Geometry of Thought*. Bradford Books, MIT Press.

Gärdenfors, Peter. 2007. Representing actions and functional properties. In *Body, Language, and Mind*, edited by Tom Ziemke, Jordan Zlatev, and Roslyn M. Frank, 167–195. Mouton de Gruyter.

Gärdenfors, Peter. 2014. *The Geometry of Meaning: Semantics Based on Conceptual Spaces*. MIT Press.

Gärdenfors, Peter. 2024a. Event structure, force dynamics and verb semantics. *Language Sciences* 102: article 101610.

Gärdenfors, Peter. 2024b. Natural concepts and the economics of cognition and communication. *Philosophia*. https://doi.org/10.1007/s11406-024-00734-4.

Gärdenfors, Peter. 2024c. Applications of convexity in semantics for natural language. *Journal of Cognitive Science* 25 (4): 431–458.

Gärdenfors, Peter, and Anders Högberg. 2017. The archaeology of teaching and the evolution of Homo docens. *Current Anthropology* 58 (2): 188–208.

Gärdenfors, Peter, and Anders Högberg. 2024. Evolution of intentional teaching. In *Oxford Handbook of Human Symbolic Evolution*, edited by Nathalie Gontier, Andy Lock and Chris Sinhg, 237–258. Oxford University Press.

Gärdenfors, Peter, Jürgen Jost, and Massimo Warglien. 2018. From actions to effects: Three constraints on event mappings. *Frontiers in Psychology* 9 (2018): 1391.

Gärdenfors, Peter, and David Makinson. 1994. Nonmonotonic inference based on expectations. *Artificial Intelligence* 65 (2): 196–245.

Gärdenfors, Peter, and Matías Osta-Vélez. 2024. Generics as expectations: Typicality and diagnosticity. *Ratio*. https://doi.org/10.1111/rati.12424.

Gärdenfors, Peter, and Andreas Stephens. 2018. Induction and knowledge-what. *European Journal for Philosophy of Science* 8 (3): 471–491.

Gärdenfors, Peter, and Massimo Warglien. 2012. Using conceptual spaces to model actions and events. *Journal of Semantics* 29 (4): 487–519.

Gärdenfors, Peter, and Frank Zenker. 2013. Theory change as dimensional change: Conceptual spaces applied to the dynamics of empirical theories. *Synthese* 190 (6): 1039–1058.

Gärdenfors, Peter, and Frank Zenker. 2010. Using conceptual spaces to model the dynamics of empirical theories. In *Belief Revision Meets Philosophy of Science*, edited by Erik Olsson and Sebastian Enqvist, 137–153. Springer Netherlands.

Garner, Wendell R. 1974. Attention: The processing of multiple sources of information. *Handbook of Perception* 2: 23–59.

Gasser, Michael, and Linda B. Smith. 1998. Learning nouns and adjectives: A connectionist account. *Language and Cognitive Processes* 13: 269–306.

Gelman, Susan A. 1988. The development of induction within natural kind and artifact categories. *Cognitive Psychology* 20 (1): 65–95.

Gelman, Susan A. 2003. *The Essential Child*. Oxford University Press.

Gelman, Susan A. 2004. Psychological essentialism in children. *Trends in Cognitive Sciences* 8 (9): 404–409.

Gelman, Susan A. 2010. Generics as a window onto young children's concepts. In *Kinds, Things, and Stuff*, edited by Francis Jeffry Pelletier, 100–120. Oxford University Press.

Gelman, Susan A. 2021. Generics in society. *Language in Society* 50 (4): 517–532.

Gelman, Susan A., and Cristine H. Kalish. 1993. Categories and causality. In *Emerging Themes in Cognitive Development*, vol. 2, edited by Robert Pasnak and Mark L. Howe. Springer-Verlag.

Gelman, Susan A., and Henry M. Wellman. 1991. Insides and essences: Early understandings of the nonobvious. *Cognition* 38: 213–244.

Gelman, Susan A., Sarah-Jane Leslie, Alexandra M. Was, and Christina M. Koch. 2015. Children's interpretations of general quantifiers, specific quantifiers and generics. *Language, Cognition and Neuroscience* 30 (4): 448–461.

Gelman, Susan A., and Steven O. Roberts. 2017. How language shapes the cultural inheritance of categories. *Proceedings of the National Academy of Sciences* 114 (30): 7900–7907.

Gentner, Dedre. 1983. Structure-mapping: A theoretical framework for analogy. *Cognitive Science* 7 (2): 155–170.

Gentner, Dedre. 2002. Analogy in scientific discovery: The case of Johannes Kepler. In *Model-Based Reasoning: Science, Technology, Values*, edited by Lorenzo Magnani and Nancy J. Nersessian, 21–39. Kluwer Academic.

Gentner, Dedre. 2003. Why we're so smart. In *Language in Mind: Advances in the Study of Language and Cognition*, edited by Dedre Gentner and Susan Goldin-Meadow, 195–235. MIT Press.

Gentner, Dedre, and Arthur B. Markman. 1997. Structure mapping in analogy and similarity. *American Psychologist* 52 (1): 45–56.

Gentner, Dedre, and Christian Hoyos. 2017. Analogy and abstraction. *Topics in Cognitive Science* 9 (3): 672–693.

Gentner, Dedre, and Keith J. Holyoak. 1997. Reasoning and learning by analogy: Introduction. *American Psychologist* 52 (1): 32–34.

Gentner, Dedre, Mary Jo Rattermann, Arthur B. Markman, and Laura Kotovsky. 1995. Two forces in the development of relational similarity. In *Developing Cognitive*

Competence: New Approaches to Process Modeling, edited by Terrence J. Simon and Graeme S. Halford, 263–313. Lawrence Erlbaum Associates.

Gibson, James J. 1979. *The Ecological Approach to Visual Perception*. Houghton Mifflin.

Giere, Ronald N. 2000. Theories. In *A Companion to the Philosophy of Science*, edited by William H. Newton-Smith, 515–524. Wiley-Blackwell.

Gigerenzer, Gerd. 1996. On narrow norms and vague heuristics: A reply to Kahneman and Tversky. *Psychological Review* 103 (3): 592–596.

Gigerenzer, Gerd, and Daniel G. Goldstein. 1996. Reasoning the fast and frugal way: Models of bounded rationality. *Psychological Review* 103 (4): 650–669.

Goldowsky, Howard, and Vasanth Sarathy. 2024. Analogical reasoning within a conceptual hyperspace. https://analogy-angle.github.io.

Goldstone, Robert L., Douglas L. Medin, and Dedre Gentner. 1991. Relational similarity and the nonindependence of features in similarity judgments. *Cognitive Psychology* 23 (2): 222–262.

Goldstone, Robert L., Douglas L. Medin, and James Halberstadt. 1997. Similarity in context. *Memory and Cognition* 25 (2): 237–255.

Goldstone, Robert L., and Lawrence W. Barsalou. 1998. Reuniting perception and conception. *Cognition* 65 (2–3): 231–262.

Goodman, Nelson. 1955. *Facts, Fiction and Forecast*. Harvard University Press.

Goodman, Nelson. 1972. Seven strictures on similarity. In *Problems and Projects*, 437–446. Bobbs-Merrill.

Gopnik, Alison, and Andrew N. Meltzoff. 1997. *Words, Thoughts, and Theories*. MIT Press.

Goswami, Usha. 1992. *Analogical Reasoning in Children*. Lawrence Erlbaum.

Goswami, Usha, and Ann L. Brown. 1990. Melting chocolate and melting snowmen: Analogical reasoning and causal relations. *Cognition* 35 (1): 69–95.

Graham, Susan A., Hayli Stock, and Annette M. E. Henderson. 2006. Nineteen-month-olds' understanding of the conventionality of object labels versus desires. *Infancy* 9: 341–350.

Greenberg, Mark, and Gilbert Harman. 2006. Conceptual role semantics. In *The Oxford Handbook of Philosophy of Language*, edited by Barry C. Smith, 295. Oxford University Press.

Grice, Paul. 1975. Logic and conversation. In *Syntax and Semantics 3: Speech Acts*, edited by Peter Cole and Jerry L. Morgan, 41–58. Academic Press.

Griggs, Richard A., and James R. Cox. 1982. The elusive thematic-materials effect in Wason's selection task. *British Journal of Psychology* 73 (3): 407–420.

Hacking, Ian. 1991. A tradition of natural kinds. *Philosophical Studies: An International Journal for Philosophy in the Analytic Tradition* 61 (1–2): 109–126.

Hacking, Ian. 2001. *An Introduction to Probability and Inductive Logic.* Cambridge University Press.

Hahn, Ulrike, and Nick Chater. 1998. Similarity and rules: Distinct? Exhaustive? Empirically distinguishable? *Cognition* 65 (2–3): 197–230.

Hampton, James A. 2007. Typicality, graded membership, and vagueness. *Cognitive Science* 31 (3): 355–384.

Hampton, James A. 2012. Generics as reflecting conceptual knowledge. *Recherches Linguistiques de Vincennes* 41: 9–24.

Hampton, James A., Alessia Passanisi, and Martin L. Jönsson. 2011. The modifier effect and property mutability. *Journal of Memory and Language* 64 (3): 233–248.

Hampton, James A., and Ian Cannon. 2004. Category-based induction: An effect of conclusion typicality. *Memory and Cognition* 32 (2): 235–243.

Han, Simon J., Keith J. Ransom, Andrew Perfors, and Charles Kemp. 2022. Human-like property induction is a challenge for large language models. In *Proceedings of the Annual Meeting of the Cognitive Science Society, 44.*

Harnad, Stevan. 1990. The symbol grounding problem. *Physica D: Nonlinear Phenomena* 42 (1–3): 335–346.

Hartmann, Stephan, and Borut Trpin. 2023. Coherence of information: What it is and why it matters. In *Proceedings of the Annual Meeting of the Cognitive Science Society* 45: 3617–3623.

Hayes, Brett K. 2007. The development of inductive reasoning. In *Inductive Reasoning: Experimental, Developmental, and Computational Approaches,* edited by Aidan Feeney and Evan Heit, 25–54. Cambridge University Press.

Hayes, Brett K., Evan Heit, and Haruka Swendsen. 2010. Inductive reasoning. *Wiley Interdisciplinary Reviews: Cognitive Science* 1 (2): 278–292.

Hayes, Brett K., John E. Taplin, and Karen I. Munro. 1996. Prior knowledge and sensitivity to feature correlations in category acquisition. *Australian Journal of Psychology* 48 (1): 27–34.

Heit, Evan. 1998. A Bayesian analysis of some forms of inductive reasoning. In *Rational Models of Cognition,* edited by Mike Oaksford and Nick Chater, 248–274. Oxford University Press.

Heit, Evan. 2000. Features of similarity and category-based induction. In *Proceedings of the Interdisciplinary Workshop on Categorization and Similarity*, 115–121. University of Edinburgh.

Heit, Evan, and Joshua Rubinstein. 1994. Similarity and property effects in inductive reasoning. *Journal of Experimental Psychology: Learning, Memory, and Cognition* 20 (2): 411–422.

Hempel, Carl G. 1965. *Aspects of Scientific Explanation, and Other Essays in the Philosophy of Science*. Free Press.

Henley, Nancy M. 1969. A psychological study of the semantics of animal terms. *Journal of Verbal Learning and Verbal Behavior* 8 (2): 176–184.

Herrnstein, Richard J. 1990. Levels of stimulus control: A functional approach. *Cognition* 37 (1): 133–166.

Heyer, Gerhard. 1990. Semantics and knowledge representation in the analysis of generic descriptions. *Journal of Semantics* 7 (1): 93–110.

Hintikka, Jaakko. 1998. What is abduction? The fundamental problem of contemporary epistemology. In *Inquiry as Inquiry: A Logic of Scientific Discovery*, 91–113. Kluwer Academic Publishers.

Hofstadter, Douglas R., and Emmanuel Sander 2013. *Surfaces and Essences: Analogy as the Fule and Fire of Thinking*. Basic books.

Holland, John H., Keith J. Holyoak, Richard E. Nisbett, and Paul R. Thagard. 1986. *Induction: Processes of Inference, Learning, and Discovery*. MIT Press.

Holyoak, Keith J. 2012. Analogy and relational reasoning. In *The Oxford Handbook of Thinking and Reasoning*, edited by Keith J. Holyoak and Robert G. Morrison, 234–259. Oxford University Press.

Holyoak, Keith J., and Paul Thagard. 1989. Analogical mapping by constraint satisfaction. *Cognitive Science* 13: 295–355.

Horty, John F. 2012. *Reasons as Defaults*. Oxford University Press.

Hout, Michael C., Megan H. Papesh, and Stephen D. Goldinger. 2013. Multidimensional scaling. *Wiley Interdisciplinary Reviews: Cognitive Science* 4 (1): 93–103.

Hume, David. 1748/1999. *An Enquiry Concerning the Human Understanding*. Edited by Tom L. Beauchamp. Oxford University Press.

Humphrey, Nicholas. 1992. *A History of the Mind: Evolution and the Birth of Consciousness*. Copernicus, Springer-Verlag.

Ibraheem, Noor A., Mokhtar M. Hasan, Rafiqul Z. Khan, and Pramod K. Mishra. 2012. Understanding color models: A review. *ARPN Journal of Science and Technology* 2 (3): 265–275.

Inhelder, Bärbel, and Jean Piaget. 1964. *The Early Growth of Logic in the Child*. Harper and Row.

Jackendoff, Ray. 1987. On Beyond Zebra: The relation of linguistic and visual information. *Cognition* 26: 80–114.

Jackendoff, Ray. 1990. On Larson's treatment of the double object construction. *Linguistic Inquiry* 21 (3): 427–456.

Jackendoff, Ray. 1992. *Semantic Structures*. MIT Press.

Jackendoff, Ray. 2012. *A User's Guide to Thought and Meaning*. Oxford University Press.

Jacquette, Dale. 2014. *Logic and How It Gets That Way*. Acumen.

Jäger, Gerhard. 2007. The evolution of convex categories. *Linguistics and Philosophy* 30:551–564.

Jani, Nilendu G., and Daniel S. Levine. 2000. A neural network theory of proportional analogy-making. *Neural Networks* 13 (2): 149–183.

Johannesson, Mikael. 2000. Modelling asymmetric similarity with prominence. *British Journal of Mathematical and Statistical Psychology* 53 (1): 121–139.

Johannesson, Mikael. 2002. Geometric models of similarity. *Lund University Cognitive Studies* 90. Lund University.

Johansson, Ingvar. 1998. Pattern as an ontological category. In *Formal Ontology in Information Systems*, edited by Nicola Guarino, 86–94. IOS Press.

Johnson, Mark. 1987. *The Body in the Mind: The Bodily Basis of Meaning, Imagination, and Reason*. University of Chicago Press.

Johnson-Laird, Philip N. 1983. *Mental Models: Towards a Cognitive Science of Language, Inference, and Consciousness*. Harvard University Press.

Johnson-Laird, Philip N. 2010. Against logical form. *Psychologica Belgica* 50 (3–4): 193–221.

Johnson-Laird, Philip N., and Ruth M. J. Byrne. 1991. *Deduction*. Lawrence Erlbaum Associates.

Jolliffe, Ian T. 2002. *Principal Component Analysis* (2nd ed.). Springer-Verlag.

Jones, Erin L., and Brian H. Ross. 2011. Classification versus inference learning contrasted with real-world categories. *Memory and Cognition* 39: 764–777.

Jones, Lawrence E., and Laura M. Koehly. 1993. Multidimensional scaling. In *A Handbook for Data Analysis in the Behavioral Sciences*, edited by Gideon Keren and Charles Lewis, 95–163. Lawrence Erlbaum Associates.

Jones, Matt, and Bradley C. Love. 2011. Bayesian fundamentalism or enlightenment? On the explanatory status and theoretical contributions of Bayesian models of cognition. *Behavioral and Brain Sciences* 34 (4): 169–188.

Jönsson, Martin L., and James A. Hampton. 2006. The inverse conjunction fallacy. *Journal of Memory and Language* 55 (3): 317–334.

Jönsson, Martin L., and James A. Hampton. 2012. The modifier effect in within-category induction: Default inheritance in complex noun phrases. *Language and Cognitive Processes* 27 (1): 90–116.

Kahneman, Daniel, and Amos Tversky. 1972. Subjective probability: A judgment of representativeness. *Cognitive Psychology* 3 (3): 430–454.

Kahneman, Daniel, and Amos Tversky. 1973. On the psychology of prediction. *Psychological Review* 80 (4): 237–251.

Kahneman, Daniel, and Dale T. Miller. 1986. Norm theory: Comparing reality to its alternatives. *Psychological Review* 93 (2): 136–153.

Kant, Immanuel. 1997. *Critique of Pure Reason*. Translated by Paul Guyer and Allen W. Wood. Cambridge University Press.

Katz, Jerrold J. 1990. *The Metaphysics of Meaning*. MIT Press.

Keil, Frank C. 1979. *Semantic and Conceptual Development: An Ontological Perspective*. Harvard University Press.

Keil, Frank C. 1989. *Concepts, Kinds, and Cognitive Development*. MIT Press.

Kerrén, Casper, Daniel Resnik, Christian Doeller, and Benjamin Griffiths. 2025. Exploring the role of dimensionality transformation in episodic memory. *Trends in Cognitive Sciences* 29 (7): 614–626.

Keßler, Carsten, Martin Raubal, and Krzysztof Janowicz. 2007. The effect of context on semantic similarity measurement. In *On the Move to Meaningful Internet Systems 2007*, edited by Robert Meersman, Zahir Tari, and Pilar Herrero. Lecture Notes in Computer Science, vol. 4806. Springer.

Kiefer, Ferenc. 1988. Linguistic, conceptual and encyclopedic knowledge: Some implications for lexicography. In *Proceedings of the 3rd EURALEX International Congress*, 1–10. Akadémiai Kiadó.

Klauer, Karl Christoph, Jochen Musch, and Birgit Naumer. 2000. On belief bias in syllogistic reasoning. *Psychological Review* 107 (4): 852–884.

Knauff, M., and L. E. Gazzo Castañeda. 2023. When nomenclature matters: Is the "new paradigm" really a new paradigm for the psychology of reasoning? *Thinking and Reasoning* 29 (3): 341–370.

Kornblith, Hilary. 1993. *Inductive Inference and Its Natural Ground: An Essay in Naturalistic Epistemology*. MIT Press.

Krifka, Manfred. 1999. Manner in dative alternation. In *West Coast Conference on Formal Linguistics* 18: 260–271.

Krifka, Manfred. 2012. Definitional generics. In *Genericity*, edited by Alda Mari, Claire Beyssade, and Fabio Del Prete, 372–389. Oxford University Press.

Krifka, Manfred, Francis Jeffry Pelletier, Gregory N. Carlson, Alice ter Meulen, Gennaro Chierchia, and Godehard Link. 1995. Genericity: An introduction. In *The Generic Book*, edited by Gregory N. Carlson and Francis Jeffry Pelletier, 1–124. University of Chicago Press.

Kruskal, Joseph B. 1964. Multidimensional scaling by optimizing goodness of fit to a nonmetric hypothesis. *Psychometrika* 29 (1): 1–27.

Kumar, Nitesh, Usashi Chatterjee, and Steven Schockaert. 2024. Ranking entities along conceptual space dimensions with LLMs: An analysis of fine-tuning strategies. arXiv preprint arXiv:2402.15337.

Kwisthout, Johan, Todd Wareham, and Iris van Rooij. 2011. Bayesian intractability is not an ailment that approximation can cure. *Cognitive Science* 35 (5): 779–784.

Lakoff, George. 1987. *Women, Fire, and Dangerous Things*. University of Chicago Press.

Lakoff, Robin. 1971. If's, and's and but's about conjunction. In *Studies in Linguistic Semantics*, edited by Charles J. Fillmore and D. Terence Langendoen, 3–114. Irvington.

Lampert, Martina. 2009. *Attention and Recombinance: A Cognitive-Semantic Investigation into Morphological Compositionality in English*. Peter Lang.

Landau, Barbara, and Ray Jackendoff. 1993. Whence and whither in spatial language and spatial cognition? *Behavioral and Brain Sciences* 16 (2): 255–265.

Langacker, Ronald W. 1987. *Foundations of Cognitive Grammar: Volume I: Theoretical Prerequisites*. Stanford University Press.

Langacker, Ronald W. 2000. *Grammar and Conceptualization*. Walter de Gruyter.

Langacker, Ronald W. 2002. *Concept, Image, and Symbol*. Walter de Gruyter.

Langacker, Ronald W. 2009. *Investigations in Cognitive Grammar*. Mouton de Gruyter.

Lascarides, Alex, and Nicholas Asher. 1993. Temporal interpretation, discourse relations and commonsense entailment. *Linguistics and Philosophy* 16 (5): 437–493.

Laver, Michael. 2014. Measuring policy positions in political space. *Annual Review of Political Science* 17 (1): 207–223.

Lawler, John M. 1973. *Studies in English Generics*. Doctoral dissertation, University of Michigan.

Lax, Eric. 2004. *The Mold in Dr. Florey's Coat: The Story of the Penicillin Miracle*. Holt Paperbacks.

LeDoux, Jospeh. 2003. The emotional brain, fear, and the amygdala. *Cellular and Molecular Neurobiology* 23: 727–738.

Lakemeyer, Gerhard, and Bernhard Nebel. 1994. Foundations of knowledge representation and reasoning. In *Foundations of Knowledge Representation and Reasoning*, edited by Gerhard Lekemeyer and Bernhard Nebel, 1–12. Springer-Verlag.

Leslie, Sarah-Jane. 2008. Generics: Cognition and acquisition. *Philosophical Review* 117 (1): 1–47.

Leslie, Sarah-Jane. 2012. Generics articulate default generalizations. *Recherches Linguistiques de Vincennes* 41: 25–44.

Leslie, Sarah-Jane, and Adam Lerner. 2016. Generic generalizations. In *Stanford Encyclopedia of Philosophy* (Winter 2016 Edition), edited by Edward N. Zalta. https://plato.stanford.edu/archives/win2016/entries/generics/.

Leslie, Sarah-Jane, Sangeet Khemlani, and Sam Glucksberg. 2011. Do all ducks lay eggs? The generic overgeneralization effect. *Journal of Memory and Language* 65 (1): 15–31.

Levin, Beth, and Malka Rappaport Hovav. 1994. *Unaccusativity: At the Syntax-Lexical Semantics Interface*. MIT Press.

Levinson, Stephen C. 2003. *Space in Language and Cognition: Explorations in Cognitive Diversity*. Cambridge University Press.

Lewis, Martha, and Jonathan Lawry. 2016. Hierarchical conceptual spaces for concept combination. *Artificial Intelligence* 237: 204–227.

Lieto, Antonio. 2021. *Cognitive Design for Artificial Minds*. Routledge.

Lieto, Antonio, Antonio Chella, and Marcello Frixione. 2017. Conceptual spaces for cognitive architectures: A lingua franca for different levels of representation. *Biologically Inspired Cognitive Architectures* 19: 1–9.

Lieto, Antonio, and Gian Luca Pozzato. 2018. A description logic of typicality for conceptual combination. In *International Symposium on Methodologies for Intelligent Systems*, 189–199. Springer.

Lieto, Antonio, and Gian Luca Pozzato. 2020. A description logic framework for commonsense conceptual combination integrating typicality, probabilities and cognitive heuristics. *Journal of Experimental and Theoretical Artificial Intelligence* 32 (5): 769–804.

Lifschitz, Vladimir, Leora Morgenstern, and David Plaisted. 2008. Knowledge representation and classical logic. In *Handbook of Knowledge Representation*, edited by Frank van Harmelen, Vladimir Lifschitz, and Bruce Porter, 3–88. Elsevier.

Lin, Pei-Jung, Paula J. Schwanenflugel, and Joseph M. Wisenbaker. 1990. Category typicality, cultural familiarity, and the development of category knowledge. *Developmental Psychology* 26 (5): 805–813.

Lindstromberg, Seth. 2010. *English Prepositions Explained*. John Benjamins Publishing Company.

Liu, Zhu, Ying Liu, Kandyang Luo, Cunlinag Kong, and Mansong Sun. 2025. Exploring the small world of word embeddings: A comparative study on conceptual spaces from LLMs of different scales. arXiv preprint arXiv:2502.11380.

Locke, John. 1689/1975. *An Essay Concerning Human Understanding*. Edited by Peter H. Nidditch. Clarendon Press.

López, Alejandro, Susan A. Gelman, Gary Gutheil, and Edward E. Smith. 1992. The development of category-based induction. *Child Development* 63 (5): 1070–1090.

Love, Bradley C. 2015. The algorithmic level is the bridge between computation and brain. *Topics in Cognitive Science* 7 (2): 230–242.

Love, Bradley C., and Steven A. Sloman. 1995. Mutability and the determinants of conceptual transformability. In *Proceedings of the Seventeenth Annual Conference of the Cognitive Science Society*, Pittsburgh, 654–659.

Lowe, E. Jonathan. 1993. Rationality, deduction and mental models. In *Rationality: Psychological and Philosophical Perspectives*, edited by Ken Manktelow and David Over, 211–230. Taylor and Frances/Routledge.

Lowet, Adam S., Chaz Firestone, and Brian J. Scholl. 2018. Seeing structure: Shape skeletons modulate perceived similarity. *Attention, Perception, and Psychophysics* 80 (5): 1278–1289.

Lu, Lijie, Jiyue Yang, Rui Shu, and Changquan Long. 2023. The default-interventionist model underlies premise typicality weakening the premise diversity effect during category-based induction: Event-related potentials evidence. *Scandinavian Journal of Psychology* 64 (3): 325–338.

MacFarlane, John G. 2000. *What Does It Mean to Say That Logic Is Formal?* Doctoral dissertation, University of Pittsburgh.

Machery, Edouard. 2009. *Doing Without Concepts*. Oxford University Press.

Macrini, Diego, Sven Dickinson, David Fleet, and Kaleem Siddiqi. 2011. Bone graphs: Medial shape parsing and abstraction. *Computer Vision and Image Understanding* 115 (7): 1044–1061.

Maddox, W. Todd. 1992. Perceptual and decisional separability. In *Multidimensional Models of Perception and Cognition*, edited by F. Gregory Ashby, 147–180. Lawrence Erlbaum.

Malt, Barbara C., and Edward E. Smith. 1984. Correlated properties in natural categories. *Journal of Verbal Learning and Verbal Behavior* 23 (2): 250–269.

Mandler, Jean M. 2004. *The Foundations of Mind: Origins of Conceptual Thought*. Oxford University Press.

Manktelow, Ken I., and Jonathan St. B. T. Evans. 1979. Facilitation of reasoning by realism: Effect or non-effect? *British Journal of Psychology* 70 (4): 477–488.

Marconi, Diego. 1997. *Lexical Competence*. MIT Press.

Marcus, Gary. 2018. Deep learning: A critical appraisal. arXiv preprint arXiv:1801.00631.

Mehrabian, Albert. 1996. Pleasure-arousal-dominance: A general framework for describing and measuring individual differences in temperament. *Current Psychology*, 14 (4): 261–292.

Markman, Arthur B., and Dedre Gentner. 2000. Structure mapping in the comparison process. *American Journal of Psychology* 113 (4): 501–538.

Markman, Ellen M. 1981. Two different principles of conceptual organization. In *Advances in Developmental Psychology*, edited by Michael E. Lamb and Ann L. Brown, 199–236. Lawrence Erlbaum.

Markman, Ellen M. 1989. *Categorization and Naming in Children: Problems of Induction*. MIT Press.

Marr, David. 1982. *Vision: A Computational Investigation into the Human Representation and Processing of Visual Information*. MIT Press.

Marr, David, and H. Keith Nishihara. 1978. Representation and recognition of the spatial organization of three-dimensional shapes. *Proceedings of the Royal Society of London. Series B. Biological Sciences* 200 (1140): 269–294.

Martin, Alex, Carol L. Wiggs, Leslie G. Ungerleider, and James V. Haxby. 1996. Neural correlates of category-specific knowledge. *Nature* 379 (6566): 649–652.

Martin, Alex, and Linda L. Chao. 2001. Semantic memory and the brain: Structure and processes. *Current Opinion in Neurobiology* 11 (2): 194–201.

Masterton, George, Frank Zenker, and Peter Gärdenfors. 2017. Using conceptual spaces to exhibit conceptual continuity through scientific change. *European Journal for Philosophy of Science* 7 (1): 127–150.

Mattos, Otávio, and Wolfram Hinzen. 2015. The linguistic roots of natural pedagogy. *Frontiers in Psychology* 6: 1424.

McGregor, Stephen, Kat Agres, Matthew Purver, and Geraint A. Wiggins. 2015. From distributional semantics to conceptual spaces: A novel computational method for concept creation. *Journal of Artificial General Intelligence* 6 (1): 55–86.

Medin, Douglas L. 1983. Structural principles in categorization. In *Perception, Cognition, and Development: Interactional Analyses*, edited by Thomas J. Tighe and Bryan E. Shepp, 203–230. Lawrence Erlbaum.

Medin, Douglas L., and Andrew Ortony. 1989. Psychological essentialism. In *Similarity and Analogical Reasoning*, edited by Stella Vosniadou and Andrew Ortony, 179–195. Cambridge University Press.

Medin, Douglas L., Elizabeth B. Lynch, and Karen O. Solomon. 2000. Are there kinds of concepts? *Annual Review of Psychology* 51 (1): 121–147.

Medin, Douglas L., John D. Coley, Gert Stroms, and Brett K. Hayes. 2003. A relevance theory of induction. *Psychonomic Bulletin and Review* 10 (3): 517–532.

Medin, Douglas L., Robert L. Goldstone, and Dedre Gentner. 1993. Respects for similarity. *Psychological Review* 100 (2): 254–278.

Medin, Douglas L., and William D. Wattenmaker. 1987. Category cohesiveness, theories, and cognitive archeology. In *Concepts and Conceptual Development: Ecological and Intellectual Factors in Categorization*, edited by Ulric Neisser, 25–62. Cambridge University Press.

Melara, Robert D. 1992. The concept of perceptual similarity: From psychophysics to cognitive psychology. In *Psychophysical Approaches to Cognition*, edited by Daniel Algom, 303–388. Elsevier.

Mercier, Hugo, and Dan Sperber, eds. 2017. *The Enigma of Reason*. Harvard University Press.

Mervis, Carolyn B., and Eleanor Rosch. 1981. Categorization of natural objects. *Annual Review of Psychology* 32: 89–115.

Mill, John Stuart. 1843. *A System of Logic*, vol 1. John W. Parker.

Minsky, Marvin. 1975. A framework for representing knowledge. In *The Psychology of Computer Vision*, edited by Patrick Henry Winston, 211–277. McGraw-Hill.

Minsky, Marvin. 1991. Logical versus analogical or symbolic versus connectionist or neat versus scruffy. *AI Magazine* 12 (2): 34–34.

Mitchell, Melanie. 2021. Abstraction and analogy-making in artificial intelligence. *Annals of the New York Academy of Sciences* 1505 (1): 79–101.

Montague, Richard. 1973. The proper treatment of quantification in ordinary English. In *Approaches to Natural Language: Proceedings of the 1970 Stanford Workshop on Grammar and Semantics*, 221–242. Springer Netherlands.

Moore, Robert C. 1982. The role of logic in knowledge representation and commonsense reasoning. AAAI-82: 428–433.

Mormann, Thomas. 2022. Topological models of columnar vagueness. *Erkenntnis* 87 (2): 693–716.

Morrison, Margaret. 2007. Where have all the theories gone? *Philosophy of Science* 74 (2): 195–228.

Moullec, Matthieu, and Igor Douven. 2025. Cheaper spaces. *Minds and Machines* 35 (1): 1–21.

Murphy, Gregory L. 2002. *The Big Book of Concepts*. MIT Press.

Murphy, Gregory L., and Douglas L. Medin. 1985. The role of theories in conceptual coherence. *Psychological Review* 92 (3): 289–316.

Nelson, Deborah G. Kelmer. 1999. Attention to functional properties in toddlers' naming and problem-solving. *Cognitive Development* 14 (1): 77–100.

Nelson, Katherine. 1996. *Language in Cognitive Development*. Cambridge University Press.

Nersessian, Nancy J. 2010. *Creating Scientific Concepts*. MIT Press.

Neufeld, Eleonore. 2022. Psychological essentialism and the structure of concepts. *Philosophy Compass* 17 (5): e12823.

Newman, George E., and Joshua Knobe. 2019. The essence of essentialism. *Mind and Language* 34 (5): 585–605.

Newman, George E., Patricia Herrmann, Karen Wynn, and Frank C. Keil. 2008. Biases towards internal features in infants' reasoning about objects. *Cognition* 107 (2): 420–432.

Nguyen, Simone P., and Tara Chevalier. 2015. Category coherence in children's inductive inferences with cross-classified entities. *Cognitive Development* 35: 137–150.

Nitsch, Alexander, Mona M. Garvert, Jacob L. S. Bellmund, Nicolas W. Schuck, and Christian F. Doeller. 2024. Grid-like entorhinal representation of an abstract value space during prospective decision making. *Nature Communications* 15 (1): 1198.

Nosofsky, Robert M. 1986. Attention, similarity, and the identification-categorization relationship. *Journal of Experimental Psychology* 115 (1): 39–57.

Nosofsky, Robert M. 1988. Exemplar-based accounts of relations between classification, recognition, and typicality. *Journal of Experimental Psychology: Learning, Memory, and Cognition* 14 (4): 700–708.

Nosofsky, Robert M. 1991. Stimulus bias, asymmetric similarity, and classification. *Cognitive Psychology* 23: 91–140.

Nosofsky, Robert M. 1992. Similarity scaling and cognitive process models. *Annual Review of Psychology* 43: 25–53.

Noyes, Alexander, and Frank C. Keil. 2019. Generics designate kinds but not always essences. *Proceedings of the National Academy of Sciences* 116 (41): 20354–20359.

Oaksford, Mike, and Nick Chater. 1991. Against logicist cognitive science. *Mind and Language* 6 (1): 1–38.

Oaksford, Mike, and Nick Chater. 1995. Theories of reasoning and the computational explanation of everyday inference. *Thinking and Reasoning* 1 (2): 121–152.

Oaksford, Mike, and Nick Chater. 2002. Commonsense reasoning, logic, and human rationality. In *Common Sense, Reasoning, and Rationality*, edited by Renee Elio, 174–214. Oxford University Press.

Oaksford, Mike, and Nick Chater. 2007. *Bayesian Rationality: The Probabilistic Approach to Human Reasoning*. Oxford University Press.

Öhman, Arne, Anders Flykt, and Francisco Esteves. 2001. Emotion drives attention: Detecting the snake in the grass. *Journal of Experimental Psychology: General* 130 (3): 466–478.

Oppenheimer, Robert. 1956. Analogy in science. *American Psychologist* 11 (3): 127–135.

Ortony, Andrew. 1979. Beyond literal similarity. *Psychological Review* 86 (3): 161–180.

Osherson, Daniel N., Edward E. Smith, Ormond Wilkie, Alejandro López, and Eldar Shafir. 1990. Category-based induction. *Psychological Review* 97 (2): 185–200.

Osherson, Daniel, Eldar Shafir, and Edward E. Smith. 1993. Ampliative inference: On choosing a probability distribution. *Cognition* 49 (3): 189–210.

Osta-Vélez, Matías, and Peter Gärdenfors. 2020. Category-based induction in conceptual spaces. *Journal of Mathematical Psychology* 96:102357.

Osta-Vélez, Matías, and Peter Gärdenfors. 2021. Nonmonotonic reasoning, expectation orderings, and conceptual spaces. *Journal of Logic, Language and Information* 31:77–97. https://doi.org/10.1007/s10849-021-09347-6.

Osta-Vélez, Matías, and Peter Gärdenfors. 2022. Analogy as a search procedure: A dimensional view. *Journal of Experimental and Theoretical Artificial Intelligence* 36 (7): 1135–1154. https://doi.org/10.1080/0952813X.2022.2125081.

Osta-Vélez, Matías, and Peter Gärdenfors. 2024. Conceptual coherence. [Preprint]. Available from: https://philsci-archive.pitt.edu/24056/.

Over, David E., and Jonathan St. B. T. Evans. 2024. *Human Reasoning*. Cambridge University Press.

Papafragou, Anna. 2015. The representation of events in language and cognition. In *The Conceptual Mind: New Directions in the Study of Concepts*, edited by Eric Margolis, 327–346. MIT Press.

Paradis, Carita. 2003. Is the notion of linguistic competence relevant in cognitive linguistics? *Annual Review of Cognitive Linguistics* 1: 247–271.

Park, Kiho, Yo Joong Choe, and Victor Veitch. 2023. The linear representation hypothesis and the geometry of large language models. arXiv preprint arXiv:2311.03658.

Partee, Barbara H. 2011. Formal semantics: Origins, issues, early impact. *Baltic International Yearbook of Cognition, Logic and Communication* 6 (1): 13.

Partee, Barbara. 1981. Montague grammar, mental representations and reality. In *Philosophy and Grammar*, edited by Sven Öhman and Stig Kanger, 59–78. Reidel.

Parthemore, Joel. 2013. The unified conceptual space theory: An enactive theory of concepts. *Adaptive Behavior* 21 (3): 168–177.

Patalano, A. L., Chin-Parker, S., and Ross, B. H. (2003). The role of coherence in category-based explanation. In *Proceedings of the 25th Annual Meeting of the Cognitive Science Society*, edited by Richard Alterman and David Kirsh, 910–915.

Patalano, Andrea L., Seth Chin-Parker, and Brian H. Ross. 2006. The importance of being coherent: Category coherence, cross-classification, and reasoning. *Journal of Memory and Language* 54 (3): 407–424.

Patel, Rohan, and Ellie Pavlick. 2002. Mapping language models to grounded conceptual spaces. In *International Conference on Learning*. https://openreview.net/forum?id=gJcEM8sxHK.

Peacocke, Christopher. 1992. *A Study of Concepts*. MIT Press.

Peirce, Charles S. 1955 [1883]. The general theory of probable inference. In *Philosophical Writings of Peirce*, edited by Justus Buchler, 190–217. Dover Publications.

Pelletier, Francis Jeffry, and Nicholas Asher. 1997. Generics and defaults. In *Handbook of Logic and Language*, edited by Johan van Benthem and Alice ter Meulen, 1125–1177. North-Holland.

Perkins, David N. 2002. Standard logic as a model of reasoning: The empirical critique. In *Studies in Logic and Practical Reasoning*, edited by Dov M. Gabbay, Ralph H. Johnson, Hans Jürgen Ohlbach, and John Woods, 187–223. Elsevier.

Piaget, Jean. 1929. *The Child's Conception of the World*. Routledge and Kegan Paul.

Piaget, Jean. 1972. Intellectual evolution from adolescence to adulthood. *Human Development* 15 (1): 1–12.

Piaget, Jean, Jacques Montangero, and Jean-Blaise Billeter. 1977. La formation des correlate. *In L'Abstraction Réfléchissante*, 115-129. Presses Universitaires de France.

Pinker, Steven. 1989. Language acquisition. In *The Foundations of Cognitive Science*, edited by Michael Posner, 359–399. MIT Press.

Poth, Nina L. 2019. Conceptual spaces, generalisation probabilities and perceptual categorisation. In *Conceptual Spaces: Elaborations and Applications*, edited by Mauri Kaipainen, Frank Zenker, Antti Hautamäki, and Peter Gärdenfors, 7–28. Springer.

Poth, Nina L. 2023. Refining the Bayesian approach to unifying generalisation. *Review of Philosophy and Psychology* 14 (3): 877–907.

Prasada, Sandeep, Sangeet Khemlani, Sarah-Jane Leslie, and Sam Glucksberg. 2013. Conceptual distinctions amongst generics. *Cognition* 126 (3): 405–422.

Pratto, Felicia, and Oliver P. John. 1991. Automatic vigilance: The attention-grabbing power of negative social information. *Journal of Personality and Social Psychology* 61 (3): 380–391.

Proffitt, Julia B., Douglas L. Medin, and John D. Coley. 2000. Expertise and category-based induction. *Journal of Experimental Psychology* 26 (4): 811–828.

Proust, Joëlle. 2015. The representational structure of feelings. In *Open Mind*, edited by Thomas Metzinger and Jennifer M. Windt, Article 25. Frankfurt am Main: MIND Group.

Quillian, M. Ross. 1967. Word concepts: A theory and simulation of some basic semantic capabilities. *Behavioral Science* 12 (5): 410–430.

Quine, Willard V. O. 1995. Naturalism; or, living within one's means. *Dialectica* 49 (2–4): 251–263.

Quine, Willard V. O. 1969. Natural kinds. In *Ontological Relativity and Other Essays*, 114–138. Columbia University Press.

Read, Stephen. 1994. Formal and material consequence. *Journal of Philosophical Logic* 23 (3): 247–265.

Regier, Terry, Paul Kay, and Naveen Khetarpal. 2007. Color naming reflects optimal partitions of color space. *Proceedings of the National Academy of Sciences* 104 (4): 1436–1441.

Regier, Terry, Paul Kay, and Richard S. Cook. 2005. Focal colors are universal after all. *Proceedings of the National Academy of Sciences* 102 (23): 8386–8391.

Rehder, Bob. 2006. When similarity and causality compete in category-based property induction. *Memory and Cognition* 34 (1): 3–16.

Rehder, Bob. 2007. Essentialism as a generative theory of classification. In *Causal Learning: Psychology, Philosophy, and Computation*, edited by Alison Gopnik and Laura Schulz, 190–207. Oxford University Press.

Rehder, Bob. 2017. Concepts as causal models: Categorization. In *The Oxford Handbook of Causal Reasoning*, edited by Michael Waldmann, 347–376. Oxford University Press.

Rehder, Bob, and Reid Hastie. 2001. Causal knowledge and categories: The effects of causal beliefs on categorization, induction, and similarity. *Journal of Experimental Psychology* 130 (3): 323–360.

Rehder, Bob, and Reid Hastie. 2004. Category coherence and category-based property induction. *Cognition* 91 (2): 113–153.

Rehder, Bob, and ShinWoo Kim. 2006. How causal knowledge affects classification: A generative theory of categorization. *Journal of Experimental Psychology: Learning, Memory, and Cognition* 32 (4): 659–683.

Rehder, Bob, and ShinWoo Kim. 2010. Causal status and coherence in causal-based categorization. *Journal of Experimental Psychology: Learning, Memory, and Cognition* 36 (5): 1171–1206.

Reiter, Raymond, and Giovanni Criscuolo. 1981. On interacting defaults. In *Proceedings 7th International Joint Conference on Artificial Intelligence (IJCAI'81)*, 270–276.

Richie, Russell, Bryan White, Sudeep Bhatia, and Michael C. Hout. 2020. The spatial arrangement method of measuring similarity can capture high-dimensional semantic structures. *Behavior Research Methods* 52: 1906–1928.

Rips, Lance J. 1975. Inductive judgments about natural categories. *Journal of Verbal Learning and Verbal Behavior* 14: 665–681.

Rips, Lance J. 1989. Similarity and the structure of concepts. In *Similarity and Analogical Reasoning*, edited by Stella Vosniadou and Andrew Ortony, 21–59. Cambridge University Press.

Rips, Lance J. 1994. *The Psychology of Proof: Deductive Reasoning in Human Thinking*. MIT Press.

Rogers, Timothy T., and James L. McClelland. 2004. *Semantic Cognition: A Parallel Distributed Processing Approach*. MIT Press.

Rosch, Eleanor. 1971. Cognitive reference points. *Cognitive Psychology* 7 (4): 532–547.

Rosch, Eleanor. 1975. Cognitive representations of semantic categories. *Journal of Experimental Psychology: General* 104 (3): 192.

Rosch, Eleanor. 1978. Prototype classification and logical classification: The two systems. In *New Trends in Cognitive Representation: Challenges to Piaget's Theory*, edited by Rochel Gelman, 73–86. Lawrence Erlbaum.

Rosch, Eleanor. 1988. Principles of categorization. In *Readings in Cognitive Science, a Perspective from Psychology and Artificial Intelligence*, edited by Allan Collins and Edward E. Smith, 312–322. Morgan Kaufmann Publishers.

Rosch, Eleanor, Carol Simpson, and R. Scott Miller. 1976. Structural bases of typicality effects. *Journal of Experimental Psychology* 2 (4): 491–502.

Rosch, Eleanor, and Carolyn B. Mervis. 1975. Family resemblances: Studies in the internal structure of categories. *Cognitive Psychology* 7 (4): 573–605.

Rosch, Eleanor, Carolyn B. Mervis, Wayne D. Gray, David M. Johnson, and Penny Boyes-Braem. 1976. Basic objects in natural categories. *Cognitive Psychology* 8 (3): 382–439.

Rumelhart, David E., and Adele A. Abrahamson. 1973. A model for analogical reasoning. *Cognitive Psychology* 5 (1): 1–28.

Russell, James. 1980. A circumplex model of affect. *Journal of Personality and Social Psychology* 39 (6): 1161–1178.

Russell, James A. 2003. Core affect and the psychological construction of emotion. *Psychological Review* 110 (1): 145–172.

Ryle, Gilbert. 1945. Knowing how and knowing that: The presidential address. In *Proceedings of the Aristotelian Society* 46: 1–16. Aristotelian Society, Wiley.

Ryle, Gilbert. 1949. *The Concept of Mind*. Hutchinson.

Sabbagh, Mark A., and Annette M. E. Henderson. 2007. How an appreciation of conventionality shapes early word learning. *New Directions for Child and Adolescent Development* 2007 (115): 25–37.

Salmon, Wesley C. 1984. *Scientific Explanation and the Causal Structure of the World*. Princeton University Press.

Schockaert, Steven, and Henri Prade. 2013. Interpolative and extrapolative reasoning in propositional theories using qualitative knowledge about conceptual spaces. *Artificial Intelligence* 202: 86–131.

Schurz, Gerhard. 1991. Relevant deduction: From solving paradoxes towards a general theory. *Erkenntnis* 35 (1): 391–437.

Schurz, Gerhard. 2014. *Philosophy of Science: A Unified Approach*. Routledge.

Schwanenflugel, Paula J., and Mario Rey. 1986. The relationship between category typicality and concept familiarity. *Memory and Cognition* 14: 150–163.

Schyns, Philippe G. 1991. A modular neural network model of concept acquisition. *Cognitive Science* 15 (4): 461–508.

Schyns, Philippe G., Robert L. Goldstone, and Jean-Pierre Thibaut. 1998. The development of features in object concepts. *Behavioral and Brain Sciences* 21 (1): 1–17.

Searle, John R. 1980. Minds, brains, and programs. *Behavioral and Brain Sciences* 3 (3): 417–424.

Sellars, Wilfrid. 1953. Inference and meaning. *Mind* 62 (247): 313–338.

Sellars, Wilfrid. 1974. Meaning as functional classification: A perspective on the relation of syntax to semantics. *Synthese* 27 (3): 417–437.

Seltman, Howard J. 2015. *Experimental Design and Analysis*. Carnegie Mellon University.

Shafir, Eldar B., Edward E. Smith, and Daniel N. Osherson. 1990. Typicality and reasoning fallacies. *Memory and Cognition* 18 (3): 229–239.

Shafir, Eldar, and Robyn A. LeBoeuf. 2002. Rationality. *Annual Review of Psychology* 53 (1): 491–517.

Shafto, Patrick, John D. Coley, and Anna Vitkin. 2007. Availability in category-based induction. In *Inductive Reasoning*, edited by Aidan Feeney and Evan Heit, 114–136. Cambridge University Press.

Shepard, Roger N. 1964. Attention and the metric structure of the stimulus space. *Journal of Mathematical Psychology* 1 (1): 54–87.

Shepard, Roger N. 1987. Toward a universal law of generalization for psychological science. *Science* 237 (4820): 1317–1323.

Shipley, Elizabeth F. 1993. Categories, hierarchies, and induction. In *Psychology of Learning and Motivation* 30:265–301. Academic Press.

Siddiqi, Kaleem, Ali Shokoufandeh, Sven J. Dickinson, and Steven W. Zucker. 1999. Shock graphs and shape matching. *International Journal of Computer Vision* 35 (1): 13–32.

Siebel, Mark. 2002. There's something about Linda: Probability, coherence and rationality. In *First Salzburg Workshop on Paradigms of Cognition*, pre-proceedings ed. by Hannes Leitgeb and Georg Schurz.

Slaughter, Virginia, and Alison Gopnik. 1996. Conceptual coherence in the child's theory of mind: Training children to understand belief. *Child Development* 67 (6): 2967–2988.

Sloman, Steven A. 1993. Feature-based induction. *Cognitive Psychology* 25: 231–280.

Sloman, Steven A. 1998. Categorical inference is not a tree: The myth of inheritance hierarchies. *Cognitive Psychology* 35 (1): 1–33.

Sloman, Steven A., Bradley C. Love, and Woo-kyoung Ahn. 1998. Feature centrality and conceptual coherence. *Cognitive Science* 22 (2): 189–228.

Sloman, Steven A., and David A. Lagnado. 2005. The problem of induction. In *Cambridge Handbook of Thinking and Reasoning*, edited by Keith J. Holyoak and Robert G. Morrison, 95–116. Cambridge University Press.

Sloutsky, Vladimir M. 2003. The role of similarity in the development of categorization. *Trends in Cognitive Sciences* 7 (6): 246–251.

Smith, Edward E., and Douglas L. Medin. 1981. *Categories and Concepts*. Harvard University Press.

Smith, Linda B. 1989. From global similarities to kinds of similarities: The construction of dimensions in development. In *Similarity and Analogical Reasoning*, edited by Stella Vosniadou and Andrew Ortony, 146–178. Cambridge University Press.

Smith, Linda B., Susan S. Jones, and Barbara Landau. 1992. Count nouns, adjectives, and perceptual properties in children's novel word interpretations. *Developmental Psychology* 28 (2): 273–286.

Solomon, Karen Olseth, and Lawrence W. Barsalou. 2004. Perceptual simulation in property verification. *Memory & Cognition* 32 (2): 244–259.

Song, Dawei, and Peter Bruza. 2001. Discovering information flow using high dimensional conceptual space. In *Proceedings of the 24th Annual International ACM SIGIR Conference on Research and Development in Information Retrieval*, 327–333.

Sowa, John F. 1991. Towards the expressive power of natural language. In *Principles of Semantic Networks*, edited by John F. Sowa, 157–189. Elsevier.

Sperber, Dan. 1994. The modularity of thought and the epidemiology of representations. In *Mapping the Mind: Domain Specificity in Cognition and Culture*, edited by Lawrence A. Hirschfeld and Susan A. Gelman, 39–67. Cambridge University Press.

Sperber, Dan, and Deirdre Wilson. 1986. *Relevance: Communication and Cognition*. Harvard University Press.

Steinert-Threlkeld, Shane, and Jakub Szymanik. 2020. Ease of learning explains semantic universals. *Cognition* 195: 104076.

Sterken, Rachel. 2015a. Generics, content and cognitive bias. *Analytic Philosophy* 56 (1): 75–93.

Sterken, Rachel. 2015b. Generics in context. *Philosopher's Imprint* 15 (21): 1–30.

Strevens, Michael. 2000. The essentialist aspect of naive theories. *Cognition* 74 (2): 149–175.

Strößner, Corina. 2022. Criteria for naturalness in conceptual spaces. *Synthese* 200 (2): 1–36.

Sznajder, Marta. 2016. What conceptual spaces can do for Carnap's late inductive logic. *Studies in History and Philosophy of Science* 56 (A): 62–71.

Taddeo, Mariarosaria, and Luciano Floridi. 2005. Solving the symbol grounding problem: A critical review of fifteen years of research. *Journal of Experimental and Theoretical Artificial Intelligence* 17 (4): 419–445.

Tajer, Diego. 2025. Logic, reasoning, and cognitive science. In *Oxford Handbook of Philosophy of Logic*, edited by Filippo Ferrari, Elke Brendel, Massimiliano Carrara, Ole Hjortland, Gil Sagi, Gila Sher, and Florian Steinberger. Oxford University Press.

Talmy, Leonard. 1983. How language structures space. In *Spatial Orientation: Theory, Research, and Application*, edited by Herbert L. Pick and Linda P. Acredolo, 225–282. Springer US.

Talmy, Leonard. 1988. Force dynamics in language and cognition. *Cognitive Science* 12 (1): 49–100.

Talmy, Leonard. 2007. Attention phenomena. In *The Oxford Handbook of Cognitive Linguistics*, edited by Dirk Geeraerts and Hubert Cuyckens, 264–294. Oxford University Press.

Tenenbaum, Joshua B., Charles Kemp, and Patrick Shafto. 2007. Theory-based Bayesian models of inductive reasoning. In *Inductive Reasoning*, edited by Aidan Feeney and Evan Heit, 167–204. Cambridge University Press.

Tenenbaum, Joshua B., and Thomas L. Griffiths. 2001. Generalization, similarity, and Bayesian inference. *Behavioral and Brain Sciences* 24 (4): 629–640.

Tenenbaum, Joshua B., Thomas L. Griffiths, and Charles Kemp. 2006. Theory-based Bayesian models of inductive learning and reasoning. *Trends in Cognitive Sciences* 10 (7): 309–318.

Tentori, Katya, Vincenzo Crupi, and Selena Russo. 2013. On the determinants of the conjunction fallacy: Probability versus inductive confirmation. *Journal of Experimental Psychology: General* 142 (1): 235–255.

Tětková, Lenka, Thea Brüsch, Teresa K. Scheidt, Fabian M. Mager, Ørtoft A. Rasmus, Jonathan Foldager, Tommy Sonne Alstrøm, and Lars Kai Hansen. 2024. On convex conceptual regions in deep network representations. *ICLR 2024 Workshop on Representational Alignment*.

Thagard, Paul. 1984. Frames, knowledge, and inference. *Synthese* 61 (2): 233–259.

Thagard, Paul. 2002. *Coherence in Thought and Action*. MIT Press.

Thagard, Paul, Chris Eliasmith, Paul Rusnock, and Cameron P. Shelley. 2002. Knowledge and coherence. In *Common Sense, Reasoning, and Rationality*, edited by Renee Elio, 104–131. Oxford University Press.

Thagard, Paul, Keith J. Holyoak, Greg Nelson, and David Gochfeld. 1990. Analog retrieval by constraint satisfaction. *Artificial Intelligence* 46 (3): 259–310.

Thagard, Paul, and Richard E. Nisbett. 1982. Variability and confirmation. *Philosophical Studies: An International Journal for Philosophy in the Analytic Tradition* 42 (3): 379–394.

Thompson, Evan. 2010. *Mind in Life: Biology, Phenomenology, and the Sciences of Mind*. Harvard University Press.

Tomlinson, Matthew T., and Bradley C. Love. 2006. Learning abstract relations through analogy to concrete exemplars. In *Proceedings of the 28th Annual Conference of the Cognitive Science Society*, 2269–2274.

Tooley, Michael. 1977. The nature of laws. *Canadian Journal of Philosophy* 7: 667–698.

Touretzky, David S. 1984. Implicit ordering of defaults in inheritance systems. In *Proceedings of the AAAI Conference*, 322–325.

Trpin, Borut and Stephan Hartmann. 2024. Explaining the conjunction fallacy. In *Proceedings of the 46th Annual Meeting of the Cognitive Science Society*, edited by Larissa Samuelson, Stefan L. Frank, Mariya Toneva, Allyson Mackey, and Eliot Hazeltine, 5220–5226.

Tulving, Endel. 1985. Memory and consciousness. *Canadian Psychology* 26 (1): 1–12.

Tulving, Endel. 2002. Episodic memory: From mind to brain. *Annual Review of Psychology* 53 (1): 1–25.

Turnbull, Margaret Greta. 2018. Underdetermination in science: What it is and why we should care. *Philosophy Compass* 13 (2): e12475.

Tversky, Amos. 1977. Features of similarity. *Psychological Review* 84 (4): 327–352.

Tversky, Amos, and Daniel Kahneman. 1974. Judgment under uncertainty: Heuristics and biases. *Science* 185 (4157): 1124–1131.

Tversky, Amos, and Daniel Kahneman. 1983. Extensional versus intuitive reasoning: The conjunction fallacy in probability judgment. *Psychological Review* 90 (4): 293.

Tyler, Andrea, and Vyvyan Evans. 2003. *The Semantics of English Prepositions: Spatial Scenes, Embodied Meaning, and Cognition*. Cambridge University Press.

Ülkümen, Gülden, Craig R. Fox, and Bertram F. Malle. 2016. Two dimensions of subjective uncertainty: Clues from natural language. *Journal of Experimental Psychology: General* 145 (10): 1280–1297.

Ungerer, Friedrich, and Hans-Jörg Schmid. 2006. *An Introduction to Cognitive Linguistics*. Pearson-Longman.

Vaina, Lucia. 1983. From shapes and movements to objects and actions: Design constraints on the representation. *Synthese* 54 (1): 3–36.

Van Rooij, Robert, and Katrin Schulz. 2020. Generics and typicality: A bounded rationality approach. *Linguistics and Philosophy* 43 (1): 83–117.

Vasilyeva, Nadya, and Tania Lombrozo. 2020. Structural thinking about social categories: Evidence from formal explanations, generics, and generalization. *Cognition* 204: 104383.

Veltman, Frank. 1998. Een zogenaamde denkfout. *Algemeen Nederlands Tijdschrift voor Wijsbegeerte* 90 (1): 11–25.

Verheyen, Steven, and Matthew Peterson. 2021. Can we use conceptual spaces to model moral principles? *Review of Philosophy and Psychology* 12: 373–395.

Verheyen, Steven, and Paul Égré. 2018. Typicality and graded membership in dimensional adjectives. *Cognitive Science* 42 (7): 2250–2286.

Vickers, John. 2014. The problem of induction. In *The Stanford Encyclopedia of Philosophy* (Fall 2014 Edition), edited by Edward N. Zalta.

Vosniadou, Stella. 1995. Analogical reasoning in cognitive development. *Metaphor and Symbol* 10 (4): 297–308.

Warglien, Massimo, and Peter Gärdenfors. 2013. Semantics, conceptual spaces, and the meeting of minds. *Synthese* 190 (12): 2165–2193.

Warglien, Massimo, Peter Gärdenfors, and Matthijs Westera. 2012. Event structure, conceptual spaces and the semantics of verbs. *Theoretical Linguistics* 38 (3–4): 159–193.

Wason, Peter C. 1968. Reasoning about a rule. *Quarterly Journal of Experimental Psychology* 20 (3): 273–281.

Wason, Peter C., and Diana Shapiro. 1971. Natural and contrived experience in a reasoning problem. *Quarterly Journal of Experimental Psychology* 23 (1): 63–71.

Wellman, Henry M., David Cross, and Julanne Watson. 2001. Meta-analysis of theory-of-mind development: The truth about false belief. *Child Development* 72 (3): 655–684.

Wheeler, Dylan, and Balasubramaniam Natarajan. 2024. Autoencoder-based domain learning for semantic communication with conceptual spaces. In *2024 Wireless Telecommunications Symposium (WTS)*, 1–6. IEEE.

Wheeler, Dylan, Erin E. Tripp, and Balasubramaniam Natarajan. 2022. Semantic communication with conceptual spaces. *IEEE Communications Letters* 27 (2): 532–535.

Wilson, Deirdre, and Dan Sperber. 2004. Relevance theory. In *The Handbook of Pragmatics*, edited by Laurence R. Horn and Gregory Ward, 607–632. Blackwell.

Winstanley, Michael A. 2022. Modelling the psychological structure of reasoning. *European Journal for Philosophy of Science* 12 (2): 1–27.

Winston, Morton E., Roger Chaffin, and Douglas Herrmann. 1987. A taxonomy of part-whole relations. *Cognitive Science* 11 (4): 417–444.

Wittgenstein, Ludwig. 2009. *Philosophical Investigations*. Wiley-Blackwell.

Wolff, Phillip. 2007. Representing causation. *Journal of Experimental Psychology: General* 136 (1): 82–111.

Wunderlich, Dieter. 1997. Cause and the structure of verbs. *Linguistic Inquiry* 28 (1): 27–68.

Yamauchi, Takashi, Bradley C. Love, and Arthur B. Markman. 2002. Learning nonlinearly separable categories by inference and classification. *Journal of Experimental Psychology: Learning, Memory, and Cognition* 28 (3): 585–593.

Yee, Eiling, Evangelia G. Chrysikou, and Sharon L. Thompson-Schill. 2014. Semantic memory. In *The Oxford Handbook of Cognitive Neuroscience: Volume 1, Core Topics*, edited by Kevin Ochsner and Stephen Kosslyn, 353–374. Oxford University Press.

Younger, Barbara. 1990. Infants' detection of correlations among feature categories. *Child Development* 61 (3): 614–620.

Zenker, Frank, and Peter Gärdenfors. 2015. Communication, rationality, and conceptual change in scientific theories. In *Applications of Conceptual Spaces: The Case for Geometric Knowledge Representation*, edited by Frank Zenker and Peter Gärdenfors, 259–277. Springer.

Zheng, Xiaochen, Martin N. Hebart, Filip Grill, Raymond J. Dolan, Christian F. Doeller, Roshan Cools, and Mona M. Garvert. 2024. Parallel cognitive maps for multiple knowledge structures in the hippocampal formation. *Cerebral Cortex* 34 (2): bhad485.

Zhu, Song Chun, and Alan L. Yuille. 1996. Forms: A flexible object recognition and modelling system. *International Journal of Computer Vision* 20 (3): 187–212.

Zimmermann, Thomas Ede. 1999. Meaning postulates and the model-theoretic approach to natural language semantics. *Linguistics and Philosophy* 22 (5): 529–561.

Zwarts, Joost, and Peter Gärdenfors. 2016. Locative and directional prepositions in conceptual spaces: The role of polar convexity. *Journal of Logic, Language and Information* 25: 109–138.

Index

Publisher contact:
The MIT Press
Massachusetts Institute of Technology
77 Massachusetts Avenue, Cambridge, MA 02139
mitpress.mit.edu

EU Authorised Representative:
Easy Access System Europe, Mustamäe tee 50,
10621 Tallinn, Estonia
gpsr.requests@easproject.com

Printed by Integrated Books International,
United States of America